THE GREENHOUSE EFFECT

THE GREENHOUSE EFFECT

Life on a Warmer Planet

Rebecca L. Johnson

Lerner Publications Company ■ Minneapolis, Minnesota

*For my readers, in the hope that they will become
guardians of the earth and its inhabitants*

This edition of this book is available in two bindings:
Library binding by Lerner Publications Company
Soft cover by First Avenue Editions
241 First Avenue North
Minneapolis, Minnesota 55401

Library of Congress Cataloging-in-Publication Data

Johnson, Rebecca L.
    The greenhouse effect: life on a warmer planet / by Rebecca L.
Johnson.
        p.   cm.
    Summary: Discusses what the greenhouse effect is, research into
its causes, and its possible impact on our planet.
    ISBN 0-8225-1591-1 ('ib. bdg.)   ISBN 0-8225-9579-6 (pbk.)
    1. Greenhouse effect, Atmospheric—Juvenile literature. 2. Global
warming—Juvenile literature. [1. Greenhouse effect, Atmospheric.
2. Global warming.] I. Title
QC912.3.J64   1990
363.73'87—dc20   89-49760
                                                         CIP

Manufactured in the United States of America

  2 3 4 5 6 7 8 9 10 99 98 97 96 95 94 93 92 91

CONTENTS

During the hot, dry summer of 1988, parched fields like this were common in the American Midwest. Some scientists think that the drought was a result of the greenhouse effect.

OUR WARMING WORLD

There is an old saying among farmers in the American Midwest that corn plants should be at least "knee-high by the Fourth of July." But in 1988, the cornfields of early July were not filled with knee-high plants. In many parts of Iowa, Illinois, Kansas, and Missouri, the corn stood only a few inches above parched soil that was as dry and hard as stone. It had not rained for weeks, and it was much hotter than usual. The withered corn plants were not just short and stunted—they were barely alive.

The summer of 1988 was one that many farmers will never forget. For it was then that the central portions of North America experienced the worst drought since the "dustbowl days" of the 1930s. Throughout the American heartland, crops were destroyed and livestock went thirsty. Deep cracks appeared in the dry soil, making normally productive farmland look like the fractured glaze on an old piece of pottery.

At the height of the drought, James E. Hansen, a climate scientist at NASA's Goddard Institute for Space Studies in New York, made a bold pronouncement. Hansen said he and several other experts were fairly certain that the terrible drought was a result of the *greenhouse effect*. Scientists have known about and have studied the greenhouse effect for many years. But up until the summer of 1988, it was not a subject familiar to most ordinary people. Hansen's well-publicized remarks changed all that. Almost overnight, the words "greenhouse effect" appeared in newspaper and magazine headlines around the world.

Hansen's comments also set off a scientific and political debate. Some people agreed with the NASA scientist, claiming that there was evidence to link the summer's hot, dry weather with the greenhouse effect. Others disagreed. They argued that the heat and drought were just part of the normal variation that usually occurs in the weather from year to year. Some believed that the drought could have been brought on by El Niño, a peculiar warming of tropical Pacific Ocean waters that takes place every three or four years. Whenever El Niño occurs (as it did in the months before the 1988 drought), it can disrupt weather patterns all over the world.

The hot, dry summer of 1988 eventually came to an end. But the debate about what caused it continues. And people everywhere are still talking about the greenhouse effect. Just what is the green-house effect? And why are so many people concerned about it?

Very simply, the **greenhouse effect** is a natural phenomenon that is responsible for trapping heat near the earth's surface and keeping the planet warm. It is called the *greenhouse* effect because the way it works is similar to the way in which a greenhouse functions.

If you have ever been inside a greenhouse at a florist's shop or garden center, the first thing you probably noticed was how much warmer it was inside the greenhouse than outside in the open air. A greenhouse operates on a fairly simple principle. Sunlight passes easily through the transparent roof and walls of the greenhouse and strikes the objects inside—plants, flowerpots, tables, and so forth. As all these things absorb energy from the sun, they become warmer. The warmed objects then give off energy themselves in the form of heat. Only a little of this heat energy passes through the glass. Much of it is "trapped" in the greenhouse, raising the temperature inside.

In some ways, the earth functions rather like a greenhouse, except on a much, much larger scale. As energy from the sun, or **solar energy**, pours down on the earth, it

Like the glass in a florist's greenhouse, the earth's atmosphere traps heat energy.

passes easily through the atmosphere that surrounds our planet, just as it does through the glass of a greenhouse. When solar energy strikes the earth, it is absorbed by the planet's surface—by the upper waters of the ocean and by the land and all the things on it. The warmed surface of the earth then radiates energy in the form of heat. Some of this heat energy passes directly into space, but some is temporarily trapped by the earth's atmosphere. Like the glass in a greenhouse, the atmosphere holds heat energy around the planet. This atmospheric heat-trapping process keeps the earth and the air just above it comfortably warm. Scientists call it the greenhouse effect.

Although the greenhouse effect has only recently been in the news, it is really nothing new in the history of our planet. In fact, it has played a key role in making life on earth possible. Without the greenhouse effect, the earth would be too cold for living things to survive. But if the greenhouse effect is natural, and even necessary for life, why are people so worried about it? Is something good turning bad?

Up until about 150 years ago, the composition of the earth's atmosphere had remained relatively unchanged for several thousand years. Since the mid-1800s, however, certain human activities have been changing the heat-trapping ability of the atmosphere dramatically. As a result of these changes, the atmosphere is now trapping more of the heat energy given off by the earth's surface than it did just a few hundred years ago.

Imagine what would happen if you sealed a greenhouse's doors, plugged all the gaps in its framework, and added an extra layer of glass. Less heat would escape from the greenhouse, and it would be warmer inside. Similarly, as the earth's atmosphere traps more of the heat energy radiating from the earth's surface, the temperature inside our "global greenhouse" goes up. Today, when most people talk about the greenhouse effect, they are referring to this increasing heat-trapping ability of the atmosphere.

Scientists who study the earth's atmosphere and climate have been anticipating this "strengthening" of the greenhouse effect for some time. They have also predicted that it will lead to **global warming**, a worldwide increase in the temperature near the earth's surface. This prediction seems to be correct because our world is definitely getting warmer. Since the beginning of this century, the average global temperature has been going up. This warming trend was especially noticeable during the 1980s. Six of the warmest years ever recorded occurred during that decade. And 1988 was the warmest of all. In that year, record high temperatures were set in many places around the world.

Is it just a coincidence that the warmest year ever recorded was also the year in which an unusually severe drought devastated much of North America? At this point, there is still no conclusive answer to this question. But one thing is clear: Global warming can be expected to change the earth's climate dramatically, and in a relatively short time. As temperatures near the earth's surface increase, most experts predict that weather patterns will be disrupted worldwide, that glaciers and polar ice caps will melt, and that sea levels will rise. Global warming could change the environment so much (and so fast) that all sorts of plants and animals might soon be in danger of extinction.

The greenhouse effect may be the most serious environmental problem the

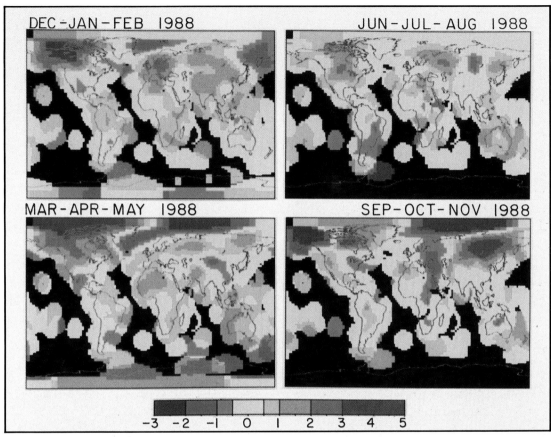

DEC-JAN-FEB 1988   JUN-JUL-AUG 1988

MAR-APR-MAY 1988   SEP-OCT-NOV 1988

-3 -2 -1 0 1 2 3 4 5

These maps produced by a computer show seasonal temperatures in 1988. The scale below indicates differences compared to average temperatures in recent years. In the areas marked with gold and red, temperatures in 1988 were as much as nine degrees Fahrenheit (five degrees Celsius) above average.

human race has ever experienced. Never before in recorded history have people been faced with the possibility of such dramatic and rapid environmental changes occurring on a global scale. For this reason, scientists around the world are working hard to achieve a better understanding of the greenhouse effect, to predict the impact it will have on earth, and to discover ways to control it. In the following pages, we will take a closer look at what scientists currently know about the greenhouse effect and how this natural process has changed in recent years.

The morning sun shines on the Florida peninsula and the Gulf of Mexico in this photograph taken from an orbiting Apollo spacecraft.

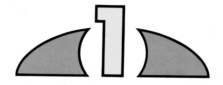

THE GLOBAL GREENHOUSE

The greenhouse effect is a complex natural process. In order to understand how it works, it is necessary to know something about the earth's atmosphere and the sun's energy. Once we have a little background information on these subjects, it will be easier to understand the way in which the atmosphere and the sun's energy interact to warm the surface of our planet.

Approximately 4.5 billion years ago, our solar system formed out of a great circling mass of gas and dust. An energy-producing star, which we call the sun, emerged at the center of the system. And a group of newly formed planets, including the earth, came to orbit around the sun.

All of the planets in our solar system are bathed in the energy given off by our sun. Those planets that are close to the sun receive more energy than those that are far away from it. Thus, the planets nearer the sun are much warmer than those that are more distant. Mercury, for example, is the planet closest to the sun. Its surface temperature can reach an incredible 648 degrees Fahrenheit (342 degrees Celsius). Farthest from the sun, Pluto has a frigid surface temperature of -369° F (-223° C).

The earth is the third planet from the sun. Its average surface temperature is a pleasant 59° F (15° C). But it is not just the distance from the sun that determines the earth's temperature. Our planet's very special atmosphere plays an important role in regulating the temperature of our world.

THE EARTH'S ATMOSPHERE

In simple terms, the atmosphere is a blanket of air that surrounds the earth. Air is a mixture of many different gases. The two that are the most plentiful are nitrogen and oxygen. About 78 percent of the volume of the atmosphere is nitrogen (symbolized N_2), and just under 21 percent is oxygen (O_2). The remaining one percent or so is made up of argon (Ar), carbon dioxide (CO_2), and very small amounts of about a dozen other gases known as **trace gases**. The atmosphere also contains water in three different forms: as tiny crystals of ice, as liquid droplets, and as invisible, gaseous **water vapor**.

Scientists often measure all the atmospheric gases except O_2 and N_2 in **parts per million (ppm)** rather than percent. For example, the concentration of carbon dioxide in the atmosphere is 0.0350 percent. But it is much easier to express this measurement as 350 parts per million, or 350 ppm. This means that in a sample of one million molecules of air, there would be 350 molecules of carbon dioxide.

The atmosphere extends many miles out from the earth's surface. But it is not a uniform layer of gas from top to bottom. Because of the pulling force of gravity on the molecules of gas that make up the air, the atmosphere is denser, or "thicker" (the gas molecules are more numerous and closer together), near the ground than it is high overhead. As you move out from the planet's surface, the atmosphere becomes progressively less dense, or "thinner," as there are fewer molecules of gas.

The atmosphere has four major layers. The boundaries between these layers are not very distinct, however, so we can say only approximately where one layer ends and the next one begins. Closest to the surface of the earth is the **troposphere**, which extends out for about 7 miles (11.3 kilometers). Air is most dense in the troposphere. The troposphere is also the most familiar part of the atmosphere. It contains the air we breathe and is the layer in which most clouds form and thunderstorms and other types of weather occur.

Above the troposphere is the second

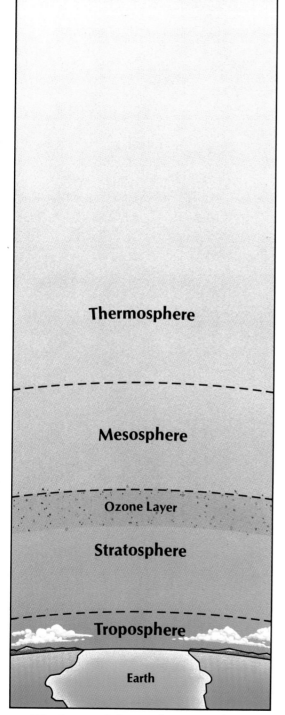

Thermosphere

Mesosphere

Ozone Layer

Stratosphere

Troposphere

Earth

The layers of the earth's atmosphere

layer, the **stratosphere**, which extends out to about 30 miles (48 km) above the planet's surface. The air in the stratosphere is much less dense than it is in the troposphere. Some types of high-flying clouds are found in this layer, but the stratosphere is best known for the fact that it contains the **ozone layer**.

Ozone is a pale blue gas. Each molecule of ozone consists of three atoms of oxygen bound together (symbolized as O_3). In high concentrations, ozone can be poisonous to humans and most other animals, and near the earth's surface, it can be a pollutant in the air. But high up in the stratosphere, ozone gas forms a protective layer that shields all living things on earth from some of the most harmful forms of radiation that come from the sun.

Above the stratosphere is the **mesosphere**, the third layer of the earth's atmosphere. The mesosphere extends to about 50 miles (80 km) above the earth. Beyond this is the fourth and final layer, the **thermosphere**. Molecules of gas in the thermosphere are few and far between. The outermost edge of the thermosphere is roughly 600 miles (965 km) above the earth's surface. Beyond this point, the airless vacuum of space begins.

THE GREENHOUSE EFFECT

As mentioned earlier, the earth's atmosphere keeps the world warm by trapping some of the heat energy emitted by the planet's surface. This greenhouse effect occurs because of the particular way that the sun's energy interacts with the atmosphere and the surface of the planet.

The sun is the ultimate source of energy for our planet. Solar energy is produced by explosive nuclear reactions that take place continuously in the interior of this fiery star. Huge amounts of solar energy constantly radiate from the sun's surface. This energy travels through space in the form of waves of **electromagnetic radiation**.

Scientists often describe waves of electromagnetic radiation in terms of their different wavelengths. What does it mean when we talk about a wave's length? If you toss a stone into a still pool of water, small waves are created that spread out from the point where the stone hit the water's surface. The up-and-down motion of the water in these waves forms a series of high points and low points. The distance from one high point to the next high point (or from one low point to the next low point) is the **wavelength** of that wave. Like water

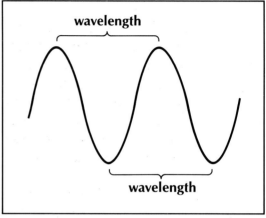

The length of a wave is measured from one high point to the next high point or from one low point to the next low point.

waves, waves of energy also have high points and low points, and their wavelengths are measured in the same way.

The waves of electromagnetic radiation that are emitted by the sun have a very wide range of wavelengths. This range is known as the **electromagnetic spectrum**. At one end of the spectrum are gamma rays, which have extremely short wavelengths. At the other end of the spectrum are radio waves, which have very long wavelengths. All the other kinds of electromagnetic radiation, each with its own characteristic wavelengths, fall in between these two extremes.

The amount of energy that a particular electromagnetic wave has depends upon its wavelength. The shorter a wave's length, the more energy that wave has. Thus, gamma rays, with the shortest wavelengths in the electromagnetic spectrum, have the greatest amount of energy. And radio waves, with the longest wavelengths, have the least.

The sun is not the only object in the universe that gives off waves of electro-magnetic radiation. All warm objects radiate some kind of energy waves. Scientists have found that the particular wavelengths that radiate from any object depend upon that object's temperature. In general, the hotter the object, the shorter the wavelengths (and the greater the energy) of the electromagnetic waves that it emits.

The surface temperature of the sun is approximately 10,800 degrees Fahrenheit (5,982° C). Most of the wavelengths given off by an object with that temperature fall between 0.2 and 3.0 micrometers (μm) in the electromagnetic spectrum (a micrometer is one millionth of a meter, or 0.000001 meter). In fact, more than 95 percent of the energy radiated by the sun

falls into this region of the spectrum. This means that although the sun gives off some gamma rays, X rays, microwaves, and radio waves, the great majority of its energy is emitted as ultraviolet radiation, visible light, and short-wave infrared radiation.

Most waves of **ultraviolet (UV) radiation** that come from the sun have wavelengths ranging from about 0.2 to 0.4 micrometers. We cannot see ultraviolet radiation. But we can sometimes see the effect of this very short-wave, high-energy radiation on our skin in the form of the sunburn that it can cause.

The wavelengths of **visible light** are somewhat longer than those of ultraviolet radiation, ranging between about 0.4 and 0.7 micrometers. Visible light is "visible" because we can see it—it is the part of the electromagnetic spectrum that includes wavelengths to which our eyes are sensi-tive. Although light coming from the sun looks white, it is actually made up of light of several colors, each with specific wavelengths. You can demonstrate this by placing a glass prism in a ray of sunshine. As the light passes through the prism, the different wavelengths can be seen as a small "rainbow" of colored light.

17

The Electromagnetic Spectrum

Gamma rays have the shortest wavelengths on the spectrum— some are less than 10-trillionth of a meter long. These extremely powerful energy waves can penetrate most substances. Even a brief exposure to gamma rays is deadly for most living things.

X rays were given their name because scientists were mystified by these powerful energy waves when they were discovered in the 1890s ("X" stands for "unknown"). Today the penetrating power of X rays is used in medicine to produce images of human bones and other internal structures on photographic film.

Ultraviolet radiation begins where X rays end and extends all the way to violet light, the shortest wavelength of visible light. Although not as powerful or penetrating as X rays, ultraviolet radiation is still dangerous to living things. Only a small percentage of UV radiation coming from the sun reaches the earth's surface, but even this small amount is enough to burn human skin and to kill small organisms such as plankton.

Visible light is the part of the electromagnetic spectrum containing wavelengths to which the eyes of humans and other animals are sensitive. Sunlight looks white but is actually made up of all the different wavelengths of the visible light spectrum, from short-wave violet light to long-wave red light. A red object appears red because it reflects the wavelengths of red light and absorbs the other wavelengths of the visible spectrum.

High Energy

Wavelength in micrometers (μm)

10^{-9} 10^{-8} 10^{-7} 10^{-6} 10^{-5} 10^{-4} 10^{-3} 10^{-2} 10^{-1} 1 10

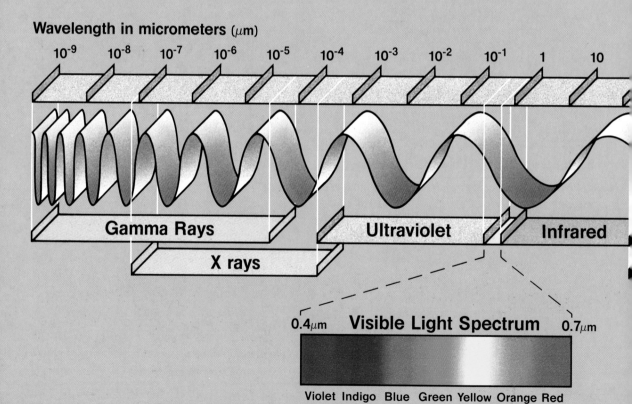

Gamma Rays

X rays

Ultraviolet

Infrared

0.4μm **Visible Light Spectrum** 0.7μm

Violet Indigo Blue Green Yellow Orange Red

Infrared radiation, lying just beyond the red end of the visible light spectrum, makes up about 60 percent of the energy given off by the sun. Humans and most other animals cannot see infrared radiation, although we can feel it as heat. Snakes have special sense organs that detect the small amounts of infrared radiation given off by warm-blooded animals. Camera film sensitive to infrared makes it possible to take photographs in places where there is no visible light.

Microwaves are very high frequency radio waves, having the shortest wavelengths of all radio waves. Microwaves were discovered in the late 1800s, but devices that could produce these energy waves did not come into use until World War II. Now microwaves play a major role in everyday life. They are used in radar, in long-distance communications, in medicine, and, of course, in microwave ovens.

Radio waves cover the greatest area on the electromagnetic spectrum. Some types have wavelengths of only a few centimeters, but at the far end of the scale, there are radio waves with wavelengths many kilometers long. The different kinds of radio waves are used primarily to carry information over great distances. Some wavelengths are used for AM/FM radio broadcasts, others for television, and still others for short-wave, mobile, and citizen-band radio.

Low Energy

10^2 10^3 10^4 10^5 10^6 10^7 10^8 10^9 10^{10} 10^{11} 10^{12} 10^{13}

Radio Waves

Microwaves

Electromagetic radiation is energy that moves through space in the form of waves. All electromagnetic radiation travels at the speed of light—186,282 miles (299,792 kilometers) per second. The major difference between each kind of radiation is its wavelength. Together, all the different wavelengths, from short-wave gamma rays to long-wave radio waves, make up the electromagnetic spectrum.

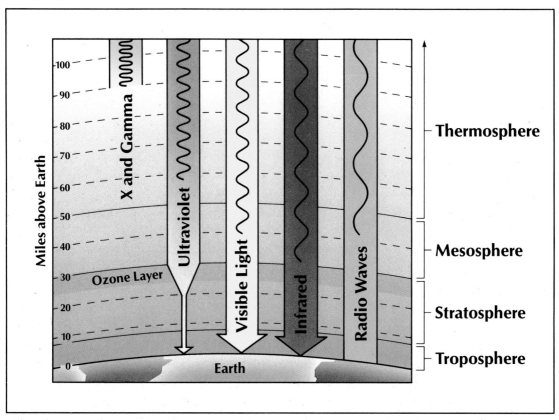

This diagram shows what happens to solar energy when it reaches the earth's atmosphere.

We cannot see **infrared radiation**, but we can feel it as heat. When you stand in the sunshine, much of the warmth you feel is infrared radiation. Waves of infrared energy coming from the sun have wavelengths ranging from approximately 0.7 to 3.0 micrometers.

Day after day, year in and year out, massive quantities of electromagnetic radiation stream out from the sun in endless waves. Most of this energy passes out of our solar system into the vast expanse of space. The rest strikes the planets, moons, and other objects that orbit the sun. What interests us here is what happens when solar energy reaches our planet, the earth.

When solar energy reaches the earth's atmosphere, a number of things happen. Most gamma rays and X rays are absorbed by gas molecules high up in the thermosphere. Almost all of the ultraviolet radiation is absorbed by the ozone layer in the stratosphere. A certain amount of solar radiation is also reflected back into space by clouds. But a large portion of the visible light and infrared radiation coming from the sun passes right through the atmosphere, without being absorbed or reflected, to strike the earth's surface. Like the glass in a greenhouse, the atmosphere is "transparent" to these energy waves.

When waves of solar energy strike the planet, some are absorbed by the earth's surface, while others are reflected back into space. Some features of the earth are highly reflective. Snow, ice, and sandy deserts tend to reflect many of the sun's rays that strike them. (That is why they appear shiny and light-colored.) On the other hand, features such as plowed fields, forests, and the oceans tend to absorb most of the energy waves that strike them.

Wherever waves of solar energy are absorbed by the earth's surface, they cause it to warm. And as it is warmed, the earth's surface radiates energy back out toward space. But what kind of energy waves are given off by the earth? Remember the rule about an object's temperature and the kind of energy it emits. The surface of the earth is much, much cooler than the surface of the sun. As a result, the energy waves that come from the warm earth have much longer wavelengths than those that radiate from the hot sun. The earth gives off mainly heat energy in the form of long-wave infrared radiation. These infrared waves have wavelengths that range from about 4 to 100 micrometers.

The fact that the energy waves emitted by the earth have much longer wavelengths than those coming from the sun is very important for the greenhouse effect. *This is because although the atmosphere is transparent to most of the sun's radiation, it is not transparent to the long-wave infrared energy coming from the earth's surface.* Instead of simply passing through the atmosphere, much of the infrared energy (heat) given off by the earth is absorbed by certain atmospheric gases.

Water vapor, carbon dioxide, and some of the trace gases all absorb part of the heat energy rising from the earth's surface. Water vapor, for example, absorbs infrared waves with wavelengths between 4 and 7 micrometers. Carbon dioxide absorbs mainly wavelengths between 13 and 100 micrometers.

Only infrared waves with wavelengths from 7 to 13 micrometers pass easily through the atmosphere and out into space. Scientists refer to this range of wavelengths as the infrared "window" because, in a sense, the atmosphere is transparent to energy waves of these lengths. This infrared energy passes out through the atmosphere as if there were nothing at all in its path.

When atmospheric gases absorb waves of infrared energy, they become warmer and give off infrared energy in return. Some of this energy heads back toward the ground, where it helps to warm the earth's surface even more. In this way, atmospheric gases "trap" heat energy coming up from the earth and prevent it from being immediately radiated out into space. This entire warming process is what scientists call the atmospheric greenhouse effect. And because the heat-trapping gases

Infrared waves with wavelengths between 7 and 13 micrometers normally pass through the earth's atmosphere and out into space.

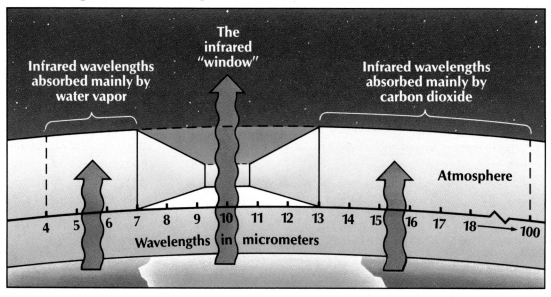

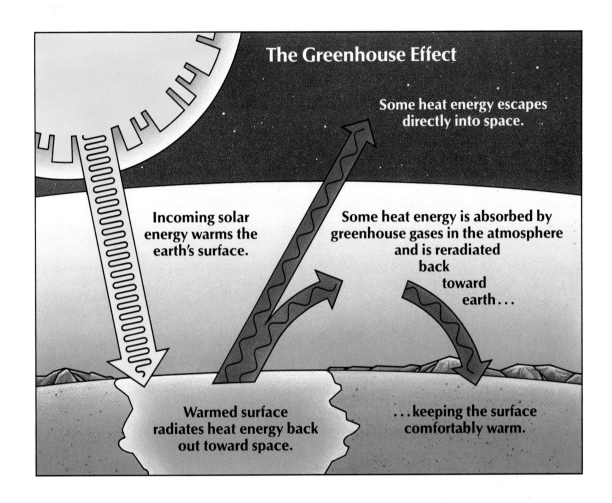

The Greenhouse Effect

Some heat energy escapes directly into space.

Incoming solar energy warms the earth's surface.

Some heat energy is absorbed by greenhouse gases in the atmosphere and is reradiated back toward earth...

Warmed surface radiates heat energy back out toward space.

...keeping the surface comfortably warm.

in the atmosphere behave like the glass of a greenhouse, they are called **greenhouse gases**.

Heat energy "trapped" by greenhouse gases is eventually radiated out into space. Thus, there is always an overall balance between the amount of incoming solar radiation and the amount of outgoing infrared radiation given off by the earth.

But because heat is trapped near the surface for a while before escaping into space, both the ground and the air just above it are kept comfortably warm.

How much does the greenhouse effect actually warm the earth? What would the earth's surface temperature be if we had no atmosphere? We can get a pretty good idea about this by comparing temperatures

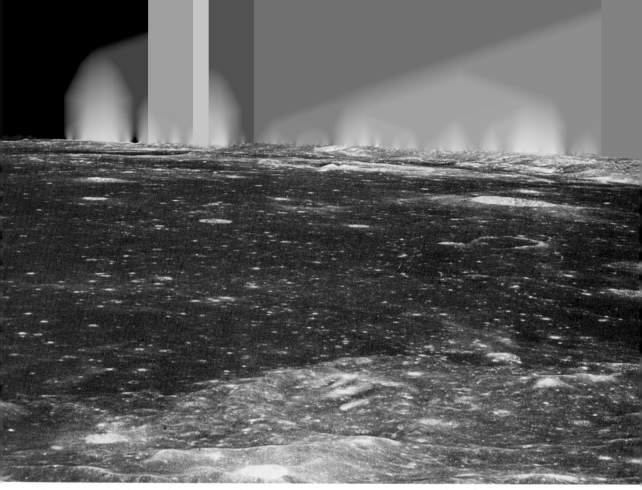

The moon and the earth are about the same distance from the sun, but because the moon has no atmosphere, it is barren and lifeless.

on the earth's surface with those on the surface of the moon. The earth and the moon are almost the same distance from the sun, so the amount of solar radiation they receive is approximately the same as well. But the moon has no atmosphere.

Temperatures on the moon alternate from very hot to very cold. During a day on the moon, for example, the surface temperature reaches a fierce 212° F (100° C). But during the night, it drops to a bitter -238° F (-150° C). The *average* temperature of the moon's surface, however, is about -13° F (-25° C). Without an atmosphere, the surface of the moon is also unprotected from the sun's harshest forms of radiation. No living thing could survive under such conditions. It is no wonder that the moon is a sterile and lifeless place.

If the earth's atmosphere were suddenly stripped away, the average temperature of the earth's surface probably would be about the same as the moon's, give or take a few degrees. But with the atmosphere in place, our world has an average temperature of about 59° F (15° C). This means that the atmosphere keeps the earth's surface 72 degrees F (40 degrees C) warmer than it would be without it!

This difference in temperature is a major reason why humans and all other living things can survive on earth. Our planet's atmosphere has made it possible for life to evolve in all its many forms, from great forests of trees to animals that inhabit every corner of the landscape to flocks of birds that soar through the sky.

You might be wondering whether just any kind of atmosphere would warm the earth in the same way that our present atmosphere does. Or is there something special about the blanket of gases that surrounds our planet? Scientists have calculated what the temperature of our world would be if we had a different kind of atmosphere, for example, one that consisted only of oxygen and nitrogen. Oxygen and nitrogen are poor greenhouse gases—they absorb very little infrared radiation. With such an atmosphere, the average surface temperature on the earth would be only a little warmer than if we had no atmosphere at all. This tells us something very important: It is the greenhouse gases in our atmosphere that are primarily responsible for keeping the planet's surface comfortably warm. Even though they are present in only small amounts, greenhouse gases have a very large effect on the temperature of our world. We would not be able to live without them.

Now that we have seen how important the greenhouse gases are, it is easy to understand that by changing the amounts of these gases in the air, we can also change the temperature near the earth's surface. Just a small increase in the concentration of any of the greenhouse gases, for example, could increase the atmosphere's capacity to absorb infrared radiation and retain heat. This, in turn, could raise temperatures worldwide.

Unfortunately, during the past 150 years or so, we humans have been doing things that have increased the atmospheric concentrations of some greenhouse gases dramatically. Carbon dioxide is one greenhouse gas that is on the rise today.

This refinery in Canada processes fuels like oil and gasoline that provide power for the machinery of modern life. Burning these fuels produces carbon dioxide, an important greenhouse gas.

2

CARBON DIOXIDE AND GLOBAL WARMING

We have already seen how carbon dioxide, along with water vapor, acts as a greenhouse gas. Obviously we need carbon dioxide to help maintain the earth's surface temperature. The problem is that there is now a lot more carbon dioxide in the air than there used to be. One important reason for this change is that for many years, we humans have been burning huge quantities of fuel (especially coal) to run our factories, heat our homes, and operate our vehicles. By burning all this fuel, we have brought about a rapid rise in the concentration of carbon dioxide in the atmosphere. Modern agricultural practices have also added carbon dioxide to the air. What exactly is carbon dioxide? And what does an increase in this gas mean in terms of the greenhouse effect?

Carbon dioxide (CO_2) is a colorless, odorless gas. A molecule of CO_2 is fairly simple, made up of just one atom of the element carbon and two atoms of oxygen. Although we cannot see it, carbon dioxide is all around us. It occurs naturally in the atmosphere. Great quantities of CO_2 are also dissolved in the world's oceans and combined with other chemicals in sediments on the ocean floor. When coal, oil, wood, or any substance containing carbon is burned, carbon dioxide is released into

27

the air. Bubbles of CO_2 are what put the "fizz" in soft drinks, beer, and champagne. Every time we exhale, we breathe out carbon dioxide. And plants cannot live and grow without this important atmospheric gas.

Carbon dioxide is constantly cycling throughout the global environment. It moves between the atmosphere and the oceans, between the atmosphere and living things, and between the soil and the atmosphere. Two important processes responsible for moving carbon dioxide from place to place on a global scale are photosynthesis and respiration.

Photosynthesis is the process by which green plants make their own food. During photosynthesis, carbon dioxide and water are combined, using energy from sunlight, to form organic (carbon-containing) compounds like sugars and starches. As a part of this process, oxygen is released— the oxygen that almost all earth's organisms must have in order to live. Land plants take up carbon dioxide for photosynthesis from the air. In the ocean, algae and photosynthesizing plankton take up CO_2 that is dissolved in seawater.

Organic compounds produced in photosynthesis are the building materials of all life on earth. For example, almost all of our food comes from plants or from animals that eat plants. Our bodies use the organic molecules in food to build and maintain cells. Organic compounds are also a rich source of energy. The energy available in organic molecules is released when they are broken down during a chemical process known as **respiration**.

Respiration takes place inside cells and is essentially the opposite of photosynthesis. In respiration, organic compounds are broken down using oxygen. During this chemical process, energy is released, and carbon dioxide and water are left as byproducts. Living things use the energy produced during respiration as a source of power for everything they do. They release the carbon dioxide into the environment. In your body, for example, the carbon dioxide produced by cellular respiration is transported to your lungs and released to the atmosphere every time you exhale.

You can see, then, that carbon dioxide is removed from the atmosphere by photosynthesizing plants. And it is returned to the atmosphere through the process of respiration. This exchange can take place quite rapidly. At the same time

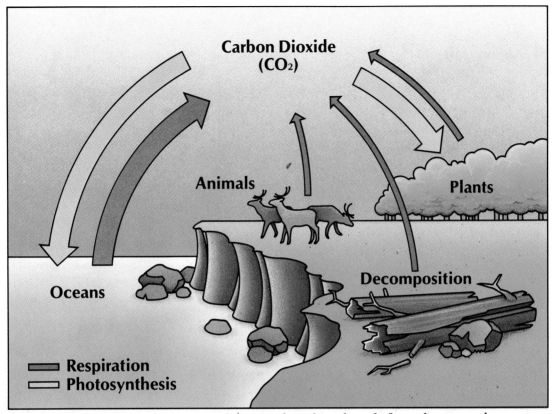

Carbon Dioxide (CO₂)

Animals

Plants

Oceans

Decomposition

▭ **Respiration**
▭ **Photosynthesis**

Photosynthesis removes approximately as much carbon dioxide from the atmosphere as respiration adds to it. This cycle keeps the atmospheric levels of CO₂ fairly constant.

as a photosynthesizing plant is making organic compounds, for example, some of the compounds are being broken down through respiration to provide the plant with energy to grow.

Most of a plant's organic compounds, however, remain as part of the plant until it dies or is eaten by an animal. If the plant dies, it eventually falls to the ground and decomposes. As it does, the plant's organic compounds are broken down through respiration by molds and bacteria in the soil. This releases carbon dioxide into the air.

What happens if the plant is eaten by an animal? Some of the plant's organic compounds are used to build the animal's body. Others pass out of the animal as waste. Like dead plant material, animal wastes decompose, and as the organic compounds in them are broken down, carbon dioxide is released into the air. In time, the animal will die, and carbon dioxide will be released into the air as its body gradually decomposes.

Thus, these two natural processes—photosynthesis and respiration—tend to be balanced in the environment. Photosynthesis removes approximately as much carbon dioxide from the atmosphere as respiration adds to it. As long as the amount of CO_2 that enters the air is the same as the amount being taken out, the concentration of carbon dioxide in the atmosphere is kept fairly constant.

CO_2 AND FOSSIL FUELS

Around 300 million years ago, during a period of the earth's history that biologists call the Carboniferous period, the climate was warmer and wetter than it is today. The concentration of carbon dioxide in the atmosphere probably was higher as well. This warm, carbon dioxide-rich environment was perfect for growing plants. Great lush forests of giant tree ferns

and other plants covered much of the land. These vast numbers of plants took up enormous quantities of CO_2 from the atmosphere and stored it in their tissues as organic compounds.

When Carboniferous plants died, some did not decompose completely. In many cases, dead plants fell into wet, swampy places where there was little oxygen in the soggy soil. When this happened, normal decomposition processes did not take place because decomposer organisms need oxygen in order to break down organic molecules. Under these conditions, plant remains accumulated year after year, layer upon layer, to form vast amounts of **peat**. After millions of years of lying buried in the ground, some of this peat was transformed into **fossil fuels**—coal, oil, and natural gas.

The massive deposits of fossil fuels that were formed during the Carboniferous period remained undisturbed beneath the earth's surface for a very long time. We have known for only a few thousand years that coal, oil, and natural gas can be burned to provide energy. It was not until the mid-1800s, however, that we began to burn very large quantities of these fossil fuels.

30

In the 1800s, a revolution swept through England, parts of Europe, and the United States. This revolution was not a military uprising, but the social and economic upheaval that we have come to call the Industrial Revolution. The Industrial Revolution was a time of rapid change in human society. Manufacturing replaced agriculture as a way of life for many people.

A large and steady supply of energy was needed to run factories and produce manufactured goods in countries that had become industrialized. This energy came from burning huge amounts of fossil fuels. And the amount of fossil fuels used in industry has been increasing ever since.

Today, besides providing energy for factories, fossil fuels are used to heat

A drawing from 1870 shows a pottery factory in Ohio with thick black coal smoke pouring from its chimneys. A coal-burning train and steamship are also adding carbon dioxide to the air.

This paper mill is one of the many modern factories that depend on fossil fuels as an energy source.

homes, offices, and schools. They power cars, trucks, trains, planes, and most other forms of transportation. Fossil fuels are also burned in power plants to produce the electricity that we use to operate everything from washing machines to computers. In other words, most of the energy needed to keep modern industrialized countries running smoothly comes from the burning of fossil fuels.

When fossil fuels are burned, the carbon dioxide that was taken out of an ancient atmosphere by plants growing during the Carboniferous period is released into our present-day atmosphere. For every ton of coal that is burned, for example, nearly 4 tons of carbon dioxide are produced and added to the air.

It has been estimated that between 1850 and 1950, approximately 60 billion tons of fossil fuel were burned, mostly in the form of coal. More recently, the worldwide consumption of fossil fuel has increased dramatically. The world now burns at least 5 billion tons of fossil fuel each year. This means that we are adding between 15 and 20 billion tons of carbon dioxide to the air every 12 months!

As this carbon dioxide from burning fossil fuels enters the atmosphere, some of it is taken up by photosynthesizing plants. Some is absorbed by the oceans. But because we are burning so much fossil fuel at such a rapid rate, we are putting carbon dioxide into the atmosphere much faster than it is being taken out by these natural processes. There is no longer a balance between the amount of carbon dioxide being added to the air and the amount of carbon dioxide being removed. As a result, the concentration of carbon dioxide in the air is steadily increasing.

DEFORESTATION

Unfortunately, burning fossil fuels is not the only thing that we humans are doing to increase the amount of carbon dioxide in the atmosphere. In many parts of the world today, forests are being destroyed at an alarming rate. Enormous numbers of trees are being cut down, both to provide timber and to clear the land for farming or ranching. This destructive process is called **deforestation**.

Deforestation is particularly severe in warm, wet tropical regions of the world where dense rain forests grow. As it happens, most of the world's rain forests also lie within the boundaries of developing countries, most of which are struggling to deal with overpopulation and poverty. Many people in these countries have no money to buy food, and no land on which to raise crops. In their search for a better life, they go into the forests and clear land to make space for farms.

In order to clear forests for agriculture, people cut down and burn all the trees in an area. When the flames die down, nothing is left but acres of blackened, lifeless countryside. The fire destroys all the plants and kills or drives off all the animals. The still-smoldering earth is then plowed up and crops are planted. In many places, the soils of rain forests are very poor, so crops can be raised on cleared land for only a few seasons. Then it becomes necessary to clear another section of the forest and start the destructive process all over again.

It is estimated that an area of tropical rain forest approximately the size of California disappears from the face of the earth each year. In Brazil alone, 20 million acres (49.5 million hectares) of the Amazon rain forest is destroyed annually. Because there has been little attempt to replant trees in deforested areas, the

Flames consume part of the Amazon rain forest in Brazil.

world's forests are disappearing very quickly. In fact, some experts predict that if deforestation continues at its current rate, all the world's rain forests could vanish within only a few decades.

Deforestation makes the problem of the greenhouse effect worse in two ways. When trees are burned, carbon dioxide is released into the air. Some researchers think that the large-scale burning of forests around the world adds at least 1 billion tons of carbon dioxide to the atmosphere each year. But deforestation does more than just add carbon dioxide to the air. It also eliminates countless numbers of CO_2-absorbing trees from the environment. As fewer and fewer trees are left to take up CO_2, the concentration of carbon dioxide in the atmosphere increases faster and faster.

MEASURING CARBON DIOXIDE

How much carbon dioxide was in the atmosphere before the Industrial Revolution began? And how much has the concentration of carbon dioxide increased since people began burning large quantities of fossil fuels and destroying the earth's

35

A team of French scientists drills an ice core at the South Pole. The air trapped in the ice provides valuable information about the earth's atmosphere in the past.

forests? Scientists have literally traveled to the ends of the earth to try and answer these questions.

For the last 25 years or so, climate scientists have been studying the ice in glaciers and ice sheets around the North and South poles. The purpose of this research is to learn more about what the earth's atmosphere and climate were like in the past. There are ice sheets in Greenland and Antarctica containing ice that is many thousands of years old. These huge sheets of ice have formed layer by layer, with the oldest ice on the bottom and the most recent ice at the top.

But how can ice tell us what conditions were like on earth long ago? Within each layer of ice are tiny bubbles of atmospheric gas that were trapped when the ice formed. Scientists get at these ancient air samples by drilling down through the ice sheets and pulling out long, cylindrical chunks of ice called **ice cores**. Once they have an ice core, they can analyze the air trapped in different parts of the core. Their analyses reveal much about what the earth's atmosphere and climate were like when each part of the ice was formed.

Air bubbles trapped in polar ice cores have provided us with an unbroken record of the earth's atmosphere that, depending on the length of the core, may extend back thousands of years. One particularly long ice core was obtained by a team of French and Soviet scientists working at Vostok Station in the Antarctic. The Vostok core, as it is called, is 7,218 feet (2,200 meters) long. It provides a record of the earth's atmosphere that goes back 160,000 years.

By analyzing air bubbles in ice cores, researchers have found that the concentration of carbon dioxide in the atmosphere has fluctuated quite a bit over the past 160,000 years. But from the time the last Ice Age ended (about 10,000 years ago) up to the mid-1800s, the concentration of carbon dioxide in the atmosphere remained fairly constant at around 270

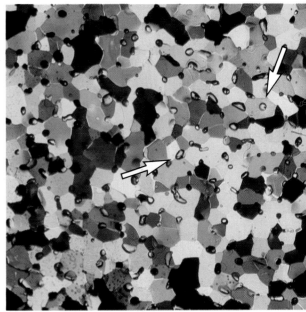

A close-up photograph of ice taken from the Antarctic Ice Sheet at a depth of about 180 feet (about 54 meters). The air bubbles trapped between the ice crystals (indicated by arrows) contain atmospheric gas from 350 years ago.

parts per million. Then, around 1850, the CO_2 concentration began to increase. This was about the same time that industrializing countries started burning large amounts of fossil fuels. And the concentration of carbon dioxide in the air has been increasing steadily ever since.

In 1957, the first systematic research began on the accumulation of carbon dioxide in the air. Charles D. Keeling of the Scripps Institution of Oceanography and his colleagues started monitoring

In 1957, Charles Keeling and his colleagues began monitoring carbon dioxide levels at this research station on the slopes of Mauna Loa in Hawaii.

carbon dioxide levels in the atmosphere. They did their work at a research station high on a mountain called Mauna Loa in the Hawaiian Islands. Mauna Loa was selected as a study site because it is in the middle of the Pacific Ocean, far from the industrial pollution of cities.

When Keeling started to keep track of carbon dioxide levels in 1957, the CO_2 concentration of the earth's atmosphere was 315 parts per million. Each year his instruments showed that the amount of carbon dioxide had increased from the year before. By 1970, it was 323 ppm.; by 1980, 335 ppm. And by the end of the 1980s, the carbon dioxide concentration in the atmosphere had risen to more than 350 ppm.

Scientists who study the atmosphere from research stations at the South Pole, on the island of Samoa, and in Alaska, Sweden, and Australia have all confirmed Keeling's findings at Mauna Loa. There is no doubt that the amount of CO_2 in the air is rising steadily and rapidly.

By burning massive amounts of fossil fuels and by destroying the earth's forests, we humans have changed the nature of the atmosphere that surrounds our planet and regulates its surface temperature. We have raised the concentration of carbon dioxide in the air from 270 ppm to 350 ppm in less than 200 years. That is an increase of 30 percent in what is really a very brief period in the earth's history.

HOW MUCH WARMER?

What does this increase mean in terms of the greenhouse effect? Is there any evidence from the past that can help us make predictions about what effect increasing levels of carbon dioxide could have on global temperature?

Many researchers are convinced that we do have such evidence. Data from the Vostok ice core, for example, show a definite relationship between carbon dioxide and global temperature. Carbon dioxide levels can be measured directly from the air bubbles in the core. By analyzing the concentration of other gases in the bubbles, scientists can also determine what the average temperature was when each of the layers of the ice sheet was formed. Using this information, it has been shown that whenever the concentration of carbon dioxide has been high over the past 160,000 years, so has the

global temperature. Likewise, whenever the concentration of CO_2 has been low, global temperatures also were lower.

Ice cores can tell us much about temperature conditions in the distant past. But what about more recent events? What has been happening to the average global temperature while carbon dioxide levels have been increasing over the past 150 years? Researchers working in England and the United States have analyzed temperature records that go back to the middle of the 1800s. They have found that since 1860, the average global temperature has risen approximately 1° F (0.6° C).

Has this recent warming trend been caused by higher levels of carbon dioxide in the air? By destroying forests and burning fossil fuels, have we raised the average temperature of our world? Some researchers say yes. They think that the recent increase in global temperature is directly related to the rise in atmospheric carbon dioxide concentration since the Industrial Revolution. Other scientists are not as certain. Although they admit that rising temperatures could be the result of rising levels of CO_2, they point out that natural variations in the earth's climate could also be responsible.

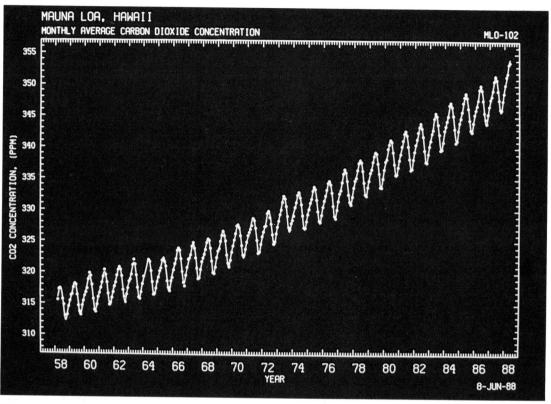

This graph shows the steady and rapid rise in carbon dioxide levels during recent years. The slight fluctuation of CO_2 that takes place each year is caused by the seasonal growth and decay of plants.

Despite doubts and disagreements about recent warming trends, however, almost all scientists agree on one point. If carbon dioxide continues to build up in the atmosphere, then more significant warming is on its way. If we continue to add carbon dioxide to the air at current rates, experts predict that by the year 2050, the concentration of atmospheric CO_2 could reach 540 ppm. That is double what it was before the Industrial Revolution began. According to most estimates, such a doubling of the atmospheric CO_2 concentration will cause the average global temperature to rise by 3.6° to 9° F (2° to 5° C).

Perhaps at this point you are wondering what all the fuss is about. How much difference could just a few degrees make? The fact is that even minor temperature changes can have a major effect on the earth's climate.

In 1815, a volcano called Tambora erupted on an island in Indonesia. Enormous amounts of ash and dust were shot up into the atmosphere, and winds distributed these small particles all around the globe. There was enough ash and dust in the air to block out some of the sun's energy and prevent it from reaching the earth's surface. During the year following Tambora's eruption, the average global temperature dropped almost a degree. And that was enough to make 1816 "the year without a summer." All around the world, it was unusually cold. In May, there were reports that wells froze in Germany. In June, hard frosts destroyed crops in New England, and in July, snow fell on the city of London, England.

Temperature changes of several degrees can have much more dramatic effects. Forty thousand years ago, the world was in the grip of the last Ice Age. Massive glaciers covered much of North America and northern Europe. Yet the average global temperature was only 9° F (5° C) colder than it is today.

As the Ice Age came to an end, temperatures gradually warmed, bringing about more dramatic changes in the earth's climate. The glaciers melted and retreated, exposing huge expanses of land. Rainfall patterns and growing seasons were altered. Because of such changes, some plant and animal species became extinct. Others flourished and spread into new areas both on land and in the sea. And all this happened largely because of an increase in the average global temperature of just a few degrees.

It now seems almost certain that increasing levels of carbon dioxide will lead to significant global warming and that the earth's climate will change as a result. But in the last few years, the problem with the greenhouse effect has become even more complex. Scientists have recently discovered that there are other greenhouse gases, in addition to CO_2, that are rapidly building up in the atmosphere. What are these gases? Where do they come from? And what effect will they have on our global greenhouse?

A farmer applying liquid nitrogen fertilizer to a corn field. The use of such fertilizers contributes to the increase of the greenhouse gas nitrous oxide in the atmosphere.

3

OTHER GREENHOUSE GASES

Until just recently, most scientists who studied the greenhouse effect were mainly interested in rising levels of CO_2 in the air. Unfortunately, the atmospheric concentrations of other less abundant, but no less important, greenhouse gases are increasing, too. Some of these gases occur naturally in the environment. Others are human-made substances that have been manufactured in laboratories. What all these gases have in common, however, is that they absorb long-wave infrared radiation in the atmosphere as well as or better than carbon dioxide does. In so doing, they trap heat near the earth's surface and contribute to the greenhouse effect. Some of these other greenhouse gases are methane, nitrous oxide, and a group of chemical substances known as chlorofluorocarbons (CFCs for short).

METHANE

After carbon dioxide, **methane** (CH_4) is the second most important greenhouse gas. There is much less methane in the air than carbon dioxide. But a molecule of methane is at least 20 times more efficient at absorbing infrared radiation than a molecule of CO_2.

In the process of digesting their food, cattle produce large amounts of methane.

Methane comes from a variety of natural sources. Cattle, for instance, produce large quantities of methane gas. The grass and other plants that cattle eat are broken down in their digestive tracts by special types of microorganisms. During digestion, great amounts of methane are produced. Cattle belch about twice every minute, and each time they do, they release methane gas into the air. Some researchers estimate that a single cow can release as much as a half pound of methane into the atmosphere each day. There may be as many as 1.3 billion cattle in the world today. They add close to 100 million tons of methane to the atmosphere every year.

Like cattle, termites depend on microorganisms located in their digestive tracts to break down the plant material that they eat. Methane gas is produced in the process. No one is yet sure how much methane these small, but incredibly numerous, insects are adding each year to the earth's atmosphere. But scientists do know that deforestation is causing the world's termite population to increase. This is because tropical rain forests that have been laid to waste are perfect breeding grounds for termites. As these insects multiply, their contribution to the greenhouse effect could become very important.

Methane is also produced by microorganisms that live in waterlogged soils, where they break down peat and other organic matter. This means that rice paddies, bogs, and swamps are all good sources of methane. In rice paddies, the slender, hollow stems of the rice plants even act as tubes through which methane gas can move up from the soggy soil and into the atmosphere. Methane also escapes

Other natural sources of methane are microorganisms in the waterlogged soil of rice paddies.

into the air from garbage dumps, coal mines, and pipes that carry natural gas (which is a mixture of methane and other gases). The burning of rainforests and grasslands adds methane to the air as well because it somehow causes soil micro-organisms to increase their methane-producing activities.

If global temperatures continue to rise, methane could become even more of a problem in the future. Perhaps the largest *potential* source of methane is the arctic tundra. The tundra is a cold, treeless plain that covers large parts of Canada and Siberia. The soils of the tundra are very rich in peat. During the brief summers that come to these regions, only the top few inches of the tundra thaw out. The rest of the soil, along with the peat it contains, remains permanently frozen. But if global warming were to melt the permanently frozen tundra soil, microorganisms would begin to break down all of the peat it contains. This would release enormous amounts of methane (as well as carbon dioxide) into the air, making the green-house effect much worse.

Researchers do not really know how much methane comes, or could come, from each of these different sources. They do know that the amount of methane in the air is increasing very rapidly, at a rate of about 1 percent each year. Currently the concentration of methane in the atmosphere is approximately 1.7 parts per million. That is nearly twice what it was just a few hundred years ago. If methane continues to accumulate in the atmosphere at the rate it is now, its concentration will double again in just 60 years.

NITROUS OXIDE

Another greenhouse gas that is on the rise is **nitrous oxide** (N_2O). Nitrous oxide (also known as laughing gas) is a colorless gas with a slightly sweet odor. It is the gas that dentists often give their patients as an anesthetic to kill pain.

Like methane, nitrous oxide comes from many sources. For example, some of the nitrous oxide in the atmosphere comes from the burning of fossil fuels, particularly coal. Automobile exhaust contains N_2O as well. Another major source is agriculture. Farmers spread nitrogen fertilizers on their fields to provide nutrients for their crops. But when these fertilizers break down in the soil, nitrous oxide is released into the air. The process of plowing up grasslands and turning them into new fields for crops also seems to add nitrous oxide to the environment.

By analyzing air trapped in ice cores, scientists have found that the atmospheric concentration of nitrous oxide was 0.28 ppm around 1900. At the end of the 1980s, it was more than 0.30 ppm. Obviously, nitrous oxide is still a minor component of the atmosphere. But it has the potential to build up very quickly. This is because nitrous oxide can remain in the troposphere for 120 to 175 years before finally rising into the stratosphere, where most of it is broken down. About 5 million tons of N_2O are added to the atmosphere each year. Since most of it remains there for more than a century, you can see how fast the nitrous oxide concentration could rise in just a few years' time.

CHLOROFLUOROCARBONS AND OTHER SYNTHETIC CHEMICALS

Chlorofluorocarbons (CFCs) are synthetic chemicals. They do not occur naturally on earth, but are manufactured in laboratories and factories. There are

Chlorofluorocarbons are used for many industrial purposes, including the manufacture of plastic foam cups and food containers.

several different kinds of CFCs. Each one contains slightly different amounts of chlorine (a gas), fluorine (another gas), and carbon. Each of the different chlorofluorocarbons is identified by a number. For example, three of the most common CFCs used until recently were CFC-11, CFC-12, and CFC-113.

Chlorofluorocarbons were discovered in the 1930s. They quickly became widely used as industrial chemicals because they are not toxic, corrosive, or flammable, and they do not react easily with other chemi-

47

cals. Today a variety of CFCs are manufactured and used in great quantities in most of the world's industrialized countries.

Some of these synthetic chemicals are used to sterilize medical equipment and to clean computer chips and electronic circuit boards. Some are used as propellants in spray cans. Others are important for the production of all sorts of plastic foam products, from soft furniture cushions to the rigid little boxes that keep fast-food hamburgers warm. CFCs also help keep our homes, offices, and schools comfortable and our food cold because they are used as coolants in air conditioners and refrigerators.

When CFCs were first put into use, they were not generally considered to be a threat to the environment, although some scientists did express concern about their safety. By the 1970s, however, it had become obvious to everyone that CFCs were definitely not harmless.

When chlorofluorocarbons are released into the environment, they rise up very slowly through the atmosphere. They can remain in the troposphere, unchanged, for 70 to 150 years. Eventually these synthetic chemicals reach the stratosphere. There they are bombarded by ultraviolet radiation coming in from the sun. Ultraviolet radiation breaks apart molecules of CFCs; in the process, atoms of chlorine are released. Chlorine atoms combine with and destroy molecules of ozone that make up the stratospheric ozone layer. A single atom of chlorine can destroy 100,000 molecules of ozone.

In 1978, because of concern about the ozone layer, CFC-11 was banned in the United States as a spray can propellant. Some other countries took the same action. However, CFC-11 was still used for other purposes. And as new CFCs were manufactured and put into use, the amounts of all these compounds in the atmosphere began to increase.

The first signs of trouble came in the early 1980s when scientists noted that the ozone layer was getting thinner, especially over the South Pole. Then, in 1985, a "hole" was discovered in the ozone layer over Antarctica. The hole, which was approximately the size of North America, appeared during the Antarctic spring. After a few months, the hole closed up again, but it has reappeared for several months each year since 1985. Recent studies show that the ozone layer is also getting very thin above the North Pole.

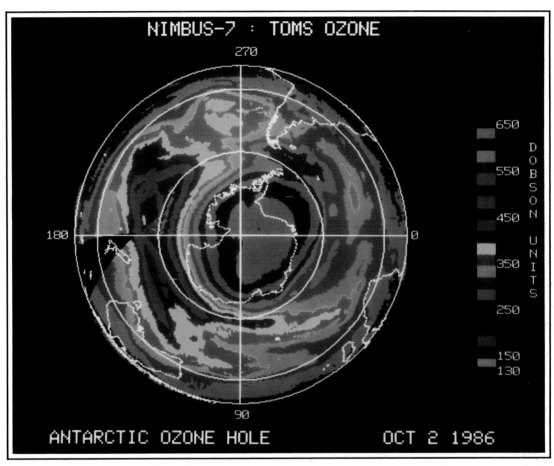

The violet color on this computer map of the Antarctic region shows the "hole" in the ozone layer that has appeared every year since 1985. The map is based on information collected by a Nimbus-7 satellite.

Holes in the ozone layer are a cause of great concern because they indicate that the earth's protective ozone layer is being damaged. Without this layer of ozone gas in the stratosphere, large amounts of harmful ultraviolet radiation would be able to reach the earth's surface. This could cause significant harm to living things and natural systems on our planet.

As it turns out, CFCs in the atmosphere are doubly dangerous. They not only destroy ozone, but they also absorb infrared radiation (heat energy) coming from the earth's surface. CFCs are very powerful greenhouse gases. One molecule of CFC-11, for example, has about the same heat-trapping effect as 10,000 molecules of carbon dioxide. And because CFCs persist in the atmosphere for so many years, they can influence the greenhouse effect for a long, long time.

Today, chlorofluorocarbons are the fastest growing of all the greenhouse gases. Atmospheric concentrations of CFC-11 and CFC-12 are increasing at a rate of about 5 percent each year. Levels of CFC-113 are growing at a frightening 11 percent every year.

And the problem keeps getting worse. Recently, scientists have discovered several other types of synthetic chemicals that act as greenhouse gases in the air. Among these is carbon tetrachloride, a chemical that was used for many years in dry cleaning clothes. Other problem chemicals are methyl chloroform, a substance used in making adhesives, and three types of **halons**, chemicals similar to chlorofluorocarbons. Because halons are very effective for putting out fires, they are widely used in fire extinguishers. The concentrations of halons and these other heat-trapping substances in the atmosphere are all increasing at an alarming rate.

"DIRTYING THE INFRARED WINDOW"

Compared to carbon dioxide, all of the other greenhouse gases that we have talked about are still only minor components of the atmosphere. Yet scientists are as concerned about the buildup of these other gases as they are about CO_2. Why is this so?

You may remember that not all the waves of infrared radiation that are given off by the earth's surface are trapped by carbon dioxide and water vapor in the atmosphere. Normally, those energy waves with wavelengths from 7 to 13 micrometers escape into space through the infrared "window."

The reason why researchers have become concerned about increases in methane, CFCs, and other greenhouse gases is that these substances *do* absorb infrared radiation with wavelengths between 7 and 13 micrometers. As a result, they trap heat energy that would normally pass into space through the "window." In a sense,

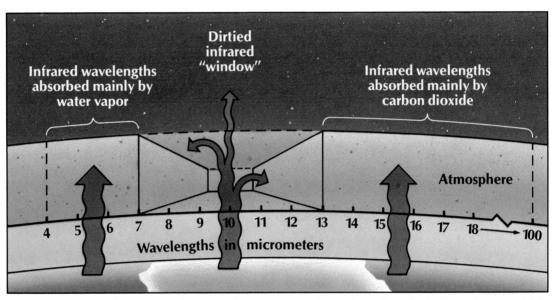

Methane, CFCs, and some other greenhouse gases "dirty the infrared window" by absorbing heat energy with wavelengths between 7 and 13 micrometers.

these gases are "dirtying the window" that has existed in our atmosphere for a very long time. As long as we continue to release them into the air, the window will get steadily "dirtier," and more and more heat will be trapped near the earth's surface. In fact, a growing number of scientists think that the combined heat-trapping effect of the other greenhouse gases already equals that of carbon dioxide.

How will this affect global warming?

Most experts predict that because of the heat-absorbing power of methane, CFCs, and other greenhouse gases, global temperatures are going to increase much faster than if carbon dioxide alone were on the rise. This means that the amount of warming predicted to take place by the middle of the next century could come years earlier.

Some questions still remain about how much global warming there will be and

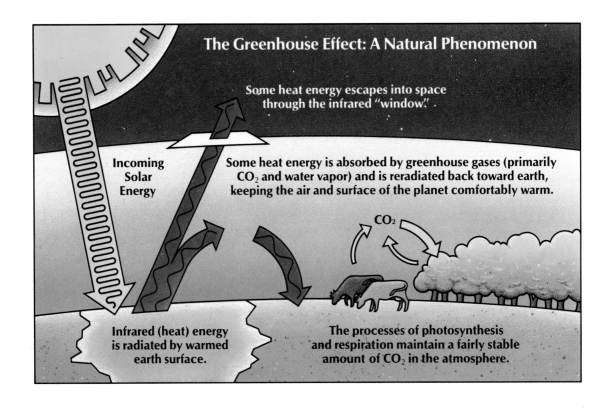

The Greenhouse Effect: A Natural Phenomenon

Some heat energy escapes into space through the infrared "window."

Incoming Solar Energy

Some heat energy is absorbed by greenhouse gases (primarily CO_2 and water vapor) and is reradiated back toward earth, keeping the air and surface of the planet comfortably warm.

CO_2

Infrared (heat) energy is radiated by warmed earth surface.

The processes of photosynthesis and respiration maintain a fairly stable amount of CO_2 in the atmosphere.

when it will take place. There is little doubt, however, that the greenhouse effect is going to be a part of our future. Nothing we can do will make the changes that have already taken place in the atmosphere disappear.

In fact, many experts on the greenhouse effect believe that the amounts of carbon

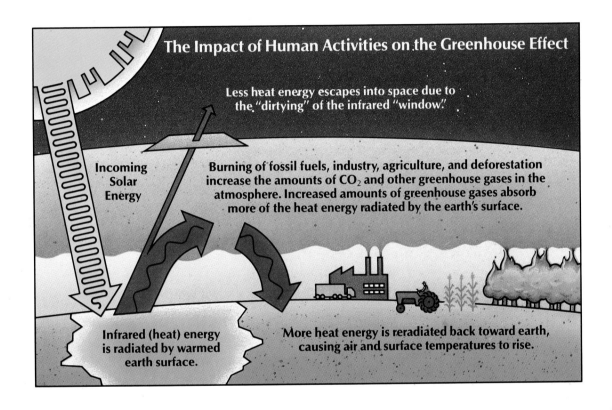

The Impact of Human Activities on the Greenhouse Effect

Less heat energy escapes into space due to the "dirtying" of the infrared "window."

Incoming Solar Energy

Burning of fossil fuels, industry, agriculture, and deforestation increase the amounts of CO$_2$ and other greenhouse gases in the atmosphere. Increased amounts of greenhouse gases absorb more of the heat energy radiated by the earth's surface.

Infrared (heat) energy is radiated by warmed earth surface.

More heat energy is reradiated back toward earth, causing air and surface temperatures to rise.

dioxide and other greenhouse gases that have been released into the air already have committed the earth to a warming of at least 1° F (0.6° C). Thus, global temperatures will warm significantly even if we could stop adding greenhouse gases to the air right now—which is something we can't do.

Zero-pressure research balloons are sent aloft to gather information about the earth's atmosphere. Scientists use this information to predict future climate changes caused by the greenhouse effect.

THE EFFECTS OF
GLOBAL WARMING

How will global warming affect our planet and the countless organisms living on it? Will warming change the earth's climate? Affect the weather? Will some parts of the world change more than others?

These are just a few of the questions being asked of scientists working on the greenhouse problem today. But finding answers to these questions is quite a challenge. The effects of global warming on the earth's climate, weather, and natural systems cannot be studied easily. You can't put weather in a test tube or investigate the oceans in a laboratory. Instead, researchers who study the greenhouse effect must rely on **climate models** to help them predict the changes that global warming could bring about on earth.

CLIMATE MODELS

Climate models are complex computer programs that can imitate the world's climate system. These programs contain vast amounts of information about the earth, especially its atmosphere and oceans. Models make it possible for scientists to duplicate, on a computer screen, the way in which the earth's climate operates in the real world. Although models are only imitations of reality, they are very useful research tools.

In order to construct a climate model, scientists must first gather as much information as possible about many different aspects of the earth. This information comes from a variety of places. Satellites orbiting far out in space send back a steady stream of data about the earth. They do such things as track ocean currents around the globe and monitor changing temperatures on land. They also provide us with photographs of the planet's surface and of the ever-changing clouds that form in the troposphere.

Instruments carried in airplanes or sent aloft with balloons also collect data about the atmosphere. This information helps scientists to understand how winds blow around the planet, how and why clouds form, and what the air temperature is in different parts of the world. Instruments on board ships or set adrift for months at sea collect data about the ocean, from its surface all the way down to its greatest depths. Information from these different sources goes into making a climate model.

Most climate models divide up the earth's surface, and the atmosphere above it, into hundreds of three-dimensional units, or "boxes." Each box is fairly large, perhaps as long as 500 miles (805 km) on a side. Using the information gathered about climate conditions on the earth's surface and in the atmosphere, scientists describe the conditions that exist inside each box with mathematical equations. There are equations to describe such things as wind, rainfall, temperature patterns, and amounts of incoming solar energy. All the boxes in the model are linked together so they can operate as a whole, as the earth's climate does.

Climate models are very complicated. It can take months to construct one that accurately represents the world's climate as it is today. But once this has been accomplished, the model is ready to help researchers predict how climate could change in response to changing conditions on the earth or in the atmosphere.

In order to do this, climate scientists program different sets of information into the model. Then they run the model on the computer and see what the outcome is. For example, a researcher might change the atmosphere in the computer program of the model by increasing the amount of methane gas that the air contains. Then she would allow the program to run.

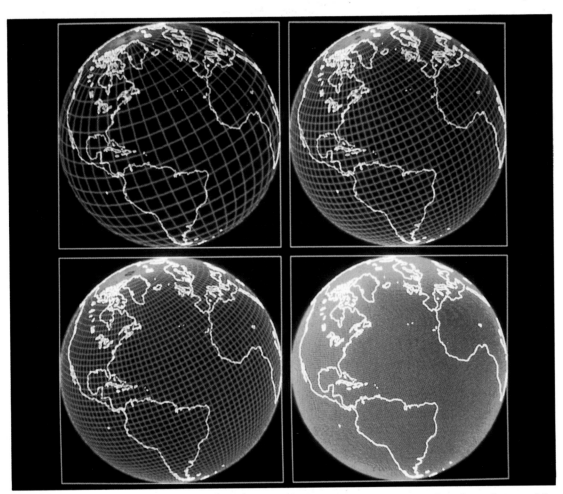

This computer-generated image of the earth pictures four of the geographical "grids" used by scientists at the National Center for Atmospheric Research. The large grid on the upper left is part of a climate-modeling program. Each box in the grid represents an area of the earth's surface and atmosphere about 500 miles (about 805 kilometers) on a side. The smaller grids are used in shorter-range weather forecasting.

Running the program can take many months while the computer makes all the necessary calculations. When the program is completed, the results would show what effects a change in methane might have on the earth's temperature, climate, and weather some time in the future.

The types of climate models most often used by greenhouse scientists are called **general circulation models**, or **GCMs**. GCMs are very detailed models. They have provided researchers with the best estimates made so far about the effect of accumulating greenhouse gases on the earth's climate. But such global climate models are far from perfect. The earth is so complex that we still have only a very limited idea about the ways in which many of its systems operate and interact. Scientists cannot include in their models what they do not know or understand.

The oceans, for instance, store and transport large amounts of heat. They also absorb and release great quantities of carbon dioxide. But scientists are only beginning to understand how important the oceans are in regulating the earth's climate.

Clouds are another problem in climate models. Clouds can affect climate in many ways. On one hand, they reflect some of the sun's energy back into space, and so help to cool the earth. On the other hand, clouds also trap some of the heat energy rising up from the earth's surface. By trapping heat, they can help to warm the earth. In order to improve the accuracy of climate models, scientists need to learn more about clouds.

Despite these problems, the climate models that greenhouse researchers use today can accurately predict how much global temperatures will increase as greenhouse gases continue to build up in the atmosphere. They also do a good job of predicting what effects global warming could have on the earth as a whole. Unfortunately, the models are not quite as reliable when it comes to making more specific predictions, such as what the effects of global warming could be in a particular country.

But as our understanding of the earth improves, so will the accuracy of climate models. Climate modeling is like putting together an enormous jigsaw puzzle. The more pieces you fit together, the more clearly you can see what the finished picture will be like. Because of the complexity of the earth, it isn't very likely that climate models will ever be 100 percent

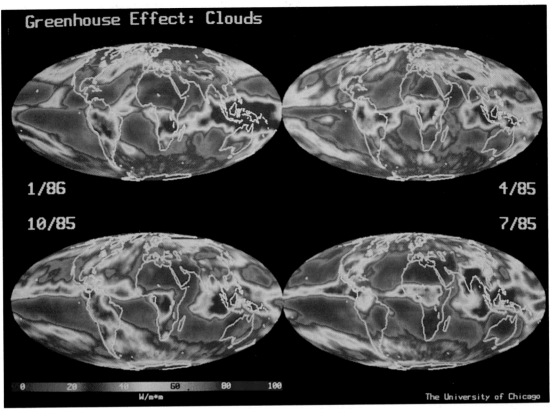

Greenhouse Effect: Clouds

1/86

4/85

10/85

7/85

0   20   40   60   80   100
W/m*m

The University of Chicago

A climate model showing how clouds affected global temperatures during four different time periods. The blue and green colors indicate areas in which clouds had a cooling effect. In the regions colored red and orange, clouds helped to trap heat and warm the earth's surface. Because of these varied effects, scientists find it difficult to predict the overall influence of clouds on global warming.

accurate—we will never be able to fit all the earth's "pieces" together. But models don't need to be perfect in order to be valuable. Despite their shortcomings, climate models like the GCMs are the best tools we have for predicting what could happen as a result of the greenhouse effect.

Let's examine the general predictions that have been made by scientists using the best climate models available. Not everyone is convinced that all these predictions will come true. But this glimpse of the future is the clearest one that scientists can give us at the moment.

July 2000

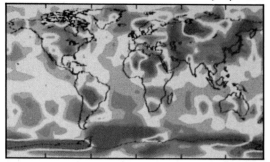

July 2015

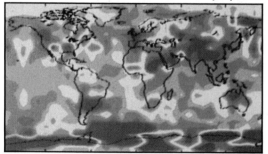

July 2029

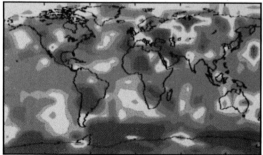

These climate-model images developed by the Goddard Institute for Space Studies show predictions for July temperature increases in selected years. The dark orange color indicates areas where average temperatures may rise as much as 9°F (5°C). These increases are expected to take place even if the present rate of CO_2 production is reduced.

A CHANGING CLIMATE

Our world could be a much different place in the 21st century. If the concentrations of heat-trapping gases in the atmosphere continue to increase, steadily rising global temperatures will cause dramatic changes in the earth's climate. In most cases, these changes will be irreversible, and their effects will be with us for a very long time. Most greenhouse researchers are fairly confident that we can expect to see the following things happen as the earth warms up.

Rising Temperatures Some parts of the world will probably warm more than others. The warming of 3.6° to 9° F (2° to 5° C) that scientists predict will occur by the middle of the next century is an increase in the *average* global temperature. Almost all climate models predict that warming will be greater at higher latitudes (nearer the poles) than at the equator. In fact, the warming that will occur at the North and South poles may be at least twice the global average. This would mean that these regions of snow and ice could warm by as much as 18° F (10° C) by the year 2050.

If global warming occurs, glaciers like this one on Baffin Island in Canada will melt and flow into the ocean, raising sea levels by many feet.

Rising Sea Levels Scientists are also certain that global warming will cause sea levels to rise. As temperatures increase worldwide, two things will happen to affect sea level. First, as the surface waters of the ocean gradually warm, the seawater will expand. This phenomenon is known as **thermal expansion**—as water is heated, it expands and takes up more space.

Researchers predict that thermal expansion alone will cause an increase in sea level of many inches. But as global temperatures continue to increase, other changes will take place. Mountain glaciers will melt, as will the great masses of ice and snow that cover Greenland and the

61

continent of Antarctica. As the ice covering these lands melts, huge quantities of water will flow into the oceans. This is expected to raise the level of the world's oceans by many feet.

The Arctic ice cap at the North Pole will also melt as global temperatures increase, but its melting will not raise sea levels. This is because Arctic ice is all floating ice that displaces, or pushes away, as much water as it contains in frozen form. You can observe the effects of displacement in a glass of water containing ice cubes. As the ice melts, the level of the water in the glass does not change.

The effects of global warming can already be seen in the world's oceans. As global temperatures have increased during this century, the level of the oceans has risen approximately 6 inches (about 15 centimeters). Most of this increase has been due to thermal expansion. Climate models indicate that by the year 2050, sea levels are likely to rise by almost 3 feet (nearly a meter). By the year 2100, the level of the sea could be more than 6 feet higher than it is now. Some models have predicted that sea levels could eventually rise much more, perhaps by as much as 15 to 24 feet (4.6 to 7.3 m)!

Changes in Precipitation We can also expect patterns of precipitation to change worldwide. On the whole, the total amount of precipitation that falls to earth each year will probably increase due to global warming. The reason for this is that as air temperatures near the ground get warmer, more water will evaporate from the planet's surface. Evaporated water eventually returns to earth as rain or snow.

An increase in the total amount of global precipitation does not mean that more rain and snow will fall everywhere on earth. Instead, some parts of the world may receive more precipitation than they do now, while others may receive less. Most global climate models predict, for example, that the interior portions of large continents such as North America and Asia will become much drier in coming years.

Changes in Plant Zones Climate change will also cause major changes in the distribution of plant life. Different kinds of plants need different kinds of climate conditions. For example, evergreen trees like those that now grow in northern Minnesota cannot grow in Oklahoma because conditions there are too hot and too dry for such trees to survive.

1990

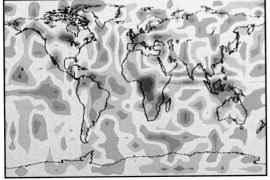

These Goddard Institute models show possible changes in patterns of precipitation during the summer months in the Northern Hemisphere. Although the total amount of precipitation will probably increase, many areas may become much drier by 2050.

2020

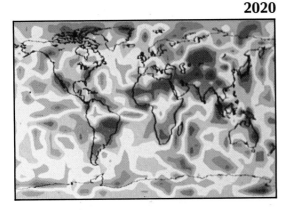

Various kinds of plants tend to grow only in specific areas, or zones, where conditions are just right to suit their needs. As global temperatures increase and climate changes, researchers expect that the various plant zones that are found in the world today will, in general, shift away from the equator and toward the poles. In the northern hemisphere, for example, the zone where evergreen trees can grow will probably move farther north than it is now.

2050

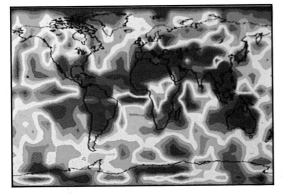

DRY · · · · · · · WET

Changes in Seasons and Violent Weather

As climate changes, so will the seasons. On the average, winters will probably be shorter and milder, and snows will melt earlier in the spring. Summers will be longer and warmer, so the growing season may be extended in some places.

Chances are good that we will see more temperature extremes in many parts of the world because of global warming. This will be especially true in the summertime, when thermometers could soar to record high temperatures much more frequently than they do now.

Global warming could cause an increase in the number and the destructive power of tornadoes and other kinds of violent weather.

It is also possible that changes in temperature could cause dangerous kinds of weather, such as tornadoes and hurricanes, to become more powerful, more destructive, and more common. The strength of hurricanes, for example, is largely determined by the temperature of the ocean waters over which these storms develop. Some atmospheric scientists think that as global warming raises water temperatures near the ocean's surface, hurricanes will have the potential to develop into even more ferocious storms than they are now.

These, then, are some of the major changes in the earth and its climate that most scientists predict as a result of the greenhouse effect and global warming. Many researchers believe that they will occur quickly, over a very short period of time. What will these dramatic and rapid changes mean for life on earth? How will climate change affect human societies? How will it affect the world's natural systems and the organisms that live in them?

In the next few pages, we will take a look at some of the more specific predictions that have been made about life on a warmer planet. As you read, it is important to keep in mind that climate models are not as accurate when it comes to making such predictions. What happens in the years to come may not be as bad as the models predict. On the other hand, it could be much worse. The value of these predictions is that they make the consequences of global warming seem much more real. They can help us understand how drastically global warming could change the world in which we live and affect all of us in the years to come.

LIFE ON A WARMER PLANET

GLOBAL WARMING AND HUMAN SOCIETIES

Changes in Agriculture

Many people are very worried about how global warming and climate change will affect agriculture. Climate affects when, where, and how the world grows its food. Most plants, including important food crops such as corn, wheat, and rice, are very sensitive to even minor changes in temperature and precipitation.

Most researchers think that the effects of global warming on agriculture will be different in different parts of the world. For example, many climate models predict that temperatures will increase and rainfall will decrease in what are now the fertile, grain-growing areas of North America, Europe, and Asia. Hotter, drier conditions could spell disaster for these regions where much of the world's food is now grown. For a while, irrigation may help to keep crops growing. But as levels of lakes, rivers, and reservoirs drop, there may not be enough water left for irrigation. In the long run, drought and summer heat could dramatically reduce food production in these regions.

In the world today, the United States and a handful of other countries usually produce much more grain than they can

A combine harvesting wheat in the midwestern United States. In the future, the production of wheat and some other grains could be greatly reduced by the effects of climate change.

use. This surplus grain is the world's insurance against famine. When countries in different parts of the world cannot produce enough food to feed their own people, they can import some of this surplus grain. But the climate changes brought about by global warming could reduce grain production enough to eliminate the grain surplus very quickly. Food shortages are considered to be one of the most serious and immediate dangers of the greenhouse effect.

It seems likely that global warming eventually will cause a permanent shift in the zones where certain types of crops can be raised around the world. For example, models show that the highly productive corn-growing region in the United States —the "corn belt"—could move several hundred miles north, where warmer temperatures will extend the growing season. This change will probably be welcomed by farmers in Minnesota and southern Canada. But it could cause serious economic problems for farmers in states such as Iowa and Illinois, where corn may no longer grow because conditions will have become too hot and dry.

On the other hand, most rice-growing regions of the world could benefit from the greenhouse effect. Rice harvests could increase in some places such as India and Southeast Asia as global warming lengthens the growing season. More land may become available for raising rice, too, as rice belts shift toward the poles. Temperatures in northern Japan, for instance, are expected to rise enough in the next 50 years so that farmers there may be able to grow about four times as much rice as they do now.

The greenhouse effect might help agriculture in another way. As you know, plants use carbon dioxide in photosynthesis. Researchers have found that when they increase the amount of carbon dioxide in the air around plants, some kinds of plants grow better and faster. Commercial growers often add CO_2 to the air inside their greenhouses in order to produce bigger, faster-growing vegetables and flowers. In the natural world, increased amounts of carbon dioxide in the atmosphere might have a similar effect. Some types of crops could possibly grow faster, develop more quickly, and produce better yields than they do now.

Unfortunately, crops are not the only plants that might grow better in a CO_2-rich environment. Many weeds could benefit in the same way. Keeping the fields clear of bigger, faster-growing weeds would mean extra work and expense for farmers. As a result, more carbon dioxide in the air might not be so good for agriculture after all.

Furthermore, carbon dioxide is just one of many things that affect plant growth. Plants also need the right temperature and enough water and nutrients to grow properly. Because of the greenhouse effect, both water shortages and higher temperatures are likely to occur in many areas of the world where crops are now grown.

Rising Sea Levels

What effect could rising sea levels have on people around the world? Rising sea levels could flood coastal cities, wash away beaches, destroy expensive property, cover valuable cropland, and drive millions of people from their homes. Of course, all these things would not happen overnight. The process would be gradual. But the higher the water rises, the more damage there will be.

What would happen, for instance, if sea levels rose by two feet (0.6 m)? Such an increase in sea level would threaten to flood many of the world's great river deltas. The broad, flat plain that lies at the mouth of Egypt's Nile River would be one of these. Many people live on the delta, and some

Low-lying islands like the Florida Keys may be completely submerged by rising sea levels.

These people in Bangladesh are building a dike to hold back the waters of a flooding river. If sea levels rise, one-fifth of the entire country might be underwater.

of the country's most productive farmland is located there. Rising sea levels would force people to abandon their homes and move inland to higher ground. The farmland would be lost to the sea.

A rise in sea level of more than two feet could cover large portions of many low-lying countries. In Bangladesh, for instance, at least one-fifth of the entire country could be underwater if sea levels rose just three feet (about 1 m). As the water advanced, millions of people would be forced to move. But where would they go? Bangladesh is a terribly overpopulated

71

country. How would millions of displaced people survive inland, where conditions are already too crowded?

If sea levels continue to rise, many islands will eventually be in danger of disappearing altogether. In the United States, the Florida Keys might not exist 60 or 70 years from now. The same thing could happen to the Maldives, a group of small islands in the Indian Ocean. None of the Maldive islands has an elevation of more than six feet (2 m). If sea levels rise by that much, as some researchers and models predict they will, the Maldives will vanish beneath the waves by the year 2100. They would be the first of several island nations to fall victim to the greenhouse effect. Further increase in sea levels would mean that many coastal cities—for example, New York, Los Angeles, London, Venice, and Shanghai—could eventually be flooded, too.

Countries that can afford to do so probably will try to protect their coastal cities from the advancing seas. Enormous amounts of sand will be needed to elevate beaches. Dikes and sea walls will have to be built, and even artificial barrier islands might need to be constructed in some places to hold back the water. Roads will have to be raised and rebuilt, and buildings raised off the ground. The cost of this battle with the sea will be enormous.

Other Effects

Because of droughts brought about by the greenhouse effect and global warming, many people could face water shortages in the future. Early-melting snows, changes in rainfall, and the need to irrigate crops when rains do not come could all limit water supplies. Rivers and wells that supply water to some cities could eventually run dry. In some parts of the world, water quality could be affected, too. Along many coasts, for example, rising seas could force salt into underground sources of drinking water.

Our health could also be affected by global warming in a rather unexpected way. As temperatures warm and climate changes, disease-carrying mosquitoes, ticks, and flies that now live only in tropical regions could invade countries at higher latitudes. In a few decades, people living in what is now the temperate zone might be worrying about contracting tropical diseases such as malaria. Insect pests that attack crops and harm farm animals could expand their ranges, too.

Forests may be severely affected by global warming and climate change.

GLOBAL WARMING AND NATURAL SYSTEMS

Global warming and climate change are likely to affect most, if not all, natural systems. Everything from mountain forests to coral reefs will probably feel the impact of the greenhouse effect. It will change the distribution of many species, disrupt natural biological communities, and probably cause some species to become extinct.

Climate changes will be especially stressful for natural systems for several

73

reasons. First, the changes will be large. Although temperature increases of a few degrees may seem small on paper, they translate into enormous changes in nature.

Second, the greenhouse effect is going to alter the environment very quickly. If climate models are correct, the earth will warm at least as much over the next 50 to 100 years as it has over the past 10,000 years, since the end of the last Ice Age. It is this very rapid rate of change that worries many biologists. In nature, rapid change is the enemy of life. The organisms that populate the earth have evolved slowly, over long periods of time. When change comes quickly, many species fail to survive.

Finally, our human-dominated landscape will make it difficult for many species to adapt to the new world brought about by the greenhouse effect. In many places, cities, roads, and farms divide up the land and isolate small populations of plants and animals. As climate changes, many living things may be trapped where they are now, unable to move to a new habitat where living conditions are right for them.

All in all, nature is going to be very hard hit by the effects of global warming and climate change. While we humans may be able to hold back rising sea levels in some areas or irrigate our fields during periods of drought, natural systems will be completely at the mercy of the greenhouse effect.

Forests

Forests are very sensitive to what happens in the environment around them. Many things affect them, from fires and insect pests to air pollution. But the two most important factors that determine where forests grow, and how well they grow, are temperature and precipitation. Because of this, researchers expect that the climate changes brought about by global warming will greatly alter the world's forests.

Most models predict that because of global warming, forests will gradually move away from the equator, toward the poles. In the Northern Hemisphere, for instance, the huge forests of evergreen trees that cover much of Canada and Asia will move northward into the arctic tundra. Where evergreens once stood, forests of deciduous trees (oak, maple, birch, and so on) will move in from the south. And where deciduous trees once grew, grasslands could eventually cover the landscape.

Forests now grow on about one-third of the land in the United States. Over the next century, however, American forests could change a great deal. Margaret Davis, an ecologist at the University of Minnesota, has used computer models to predict how four species of trees now growing in the U.S. will respond to global warming. The models predict that, during the next century, all four tree species will need to move hundreds of miles north to find the right conditions for growth.

Beech trees, for example, now grow in the eastern United States from the Canadian border all the way down to Florida. If global warming progresses as most scientists think it will, beech forests will gradually die out in most of this area. A century from now, the type of habitat in which beech trees can grow will be found only in the northeastern corner of the country.

In thinking about forests and how they might change in the years to come, it is important to keep in mind that forests are complex natural systems, not simply groups of trees. Every forest contains a unique collection of plants and animals. As forests spread into new areas, they will

In the future, deciduous trees like these may grow in areas once covered by evergreen forests.

probably undergo drastic changes in the process. Some kinds of trees may die out. So might some of the shrubs and other plants that grow near the forest floor. When the plants in a forest change, so do the animals that inhabit it. Some forest creatures may not be able to cope with what is happening to their world, and may become extinct.

We could begin to see the effects of global warming on the earth's forests in as little as 20 or 30 years. As temperatures rise, trees will begin to die out along the southern edges of the zones in which they are now found. Once trees start dying, forests could change very quickly. But no one is sure just how much they will change. At this point, the only thing that scientists can tell us with any certainty is that, because of the greenhouse effect, the forests of the future will be very different than they are today.

The Arctic

Almost all climate models predict that global warming will be greater at higher latitudes than it will be at the equator. This means that plants and animals living near the North and South poles will be especially hard hit by the greenhouse effect.

The Arctic may be one of the first natural areas on earth to show signs of damage from global warming. As temperatures go up, the polar ice cap will start to melt. Many animals live on the great sheets of sea ice that surround the North Pole. Walruses and many kinds of seals, for example, use the ice as a platform for fishing. The ice is also a place where spotted seals and harp seals come to breed and raise their pups. Polar bears, perhaps the most familiar of all arctic animals, migrate long distances across the ice in search of food and mates.

Beneath the sea ice is a less familiar arctic community. Tiny algae grow on the underside of the ice and in the cold seawater beneath it. Algae provide food for some kinds of plankton, which, in turn, are food for fish, seabirds, seals, and whales.

The arctic ice cap is already shrinking in size. Satellite pictures taken of the North Pole show that the sea ice has been decreasing for about 15 years. If temperatures warm enough, the ice could disappear completely sometime in the next century. When it does, so will most of the plants and animals that depend on it. Furthermore, once the ice is gone,

This female seal and her pup are among the many animals that live on the sea ice surrounding the North Pole. What will happen to these animals if global warming causes the ice to melt?

seawater will be able to flow freely around the North Pole. Any change in ocean currents could easily change the way in which marine life is distributed in the Arctic Ocean and possibly in other parts of the world as well.

South of the ice cap at the top of our world lies the arctic tundra, a vast plain that is frozen and snow-covered for most of the year. For several months each summer, however, the top few inches of the frozen soil melt. Small plants grow up

and quickly cover every bit of ground. Thousands of shallow ponds form, and in these ponds all sorts of insects and other tiny creatures breed and develop. They provide food for flocks of migratory birds that come to the tundra each year to nest and raise their chicks.

Because of global warming, most of the tundra could be gone in less than a hundred years. As global temperatures increase, the frozen soil will gradually thaw. Beginning at the tundra's southern

If the frozen soil of the tundra thaws, this fragile environment could be permanently changed.

border, forests of evergreen trees will move in and slowly spread northward. Many of the plants and animals that now live on the tundra will be forced to migrate farther north as their environment changes. If most climate models are correct, we can expect that in the next century only a few small patches of tundra will be left along the northernmost edges of North America and Asia.

Sadly, it looks as if the tundra is already starting to thaw. Researchers in Canada have discovered that during the past few summers, more of the frozen soil has melted than ever before. Could this be the beginning of the end for the tundra? Many people also worry that as the tundra thaws, the vast amounts of peat it contains will decompose. As a result, huge amounts of carbon dioxide and methane could be released into the atmosphere during the next few decades. Adding more of these heat-trapping gases to the air can only make the greenhouse effect much worse.

The Oceans

There is little doubt that the oceans will be affected by the greenhouse effect. But it is hard to predict just what those effects will be. This is because we know much less about the sea and its inhabitants than we do about the land and the organisms living on it. Some researchers, however, have speculated about what the greenhouse effect could mean for the marine environment.

Because of global warming, the temperature of the oceans' surface waters will increase. This change could affect the flow of ocean water around the globe and the distribution of marine life. In the future, for example, many commercially important types of fish may be found nearer the poles than they are now. Finding and catching these fish could be difficult and expensive for people who make their living from the sea.

Many marine animals such as sea urchins and starfish produce tiny offspring (larvae) that often drift great distances with ocean currents. If the currents carry the larvae to places where conditions are just right, they can settle down on the ocean floor and grow into adults. If ocean currents change because of global warming, the larvae of some species could be swept into parts of the sea where they cannot survive. Eventually these species could become extinct.

Warmer ocean waters could spell trouble for coral reefs. Corals are tiny animals that live together in colonies. They are found mainly in the tropical parts of the world. Each coral individual secretes a stony skeleton around itself that remains after the animal dies. New generations of corals live on top of the skeletons of previous generations, and over time, huge masses of these skeletons form coral reefs. Coral reefs teem with life—nearly one-third of all tropical fish species live around coral reefs.

Most kinds of reef-building corals have algae in their bodies that carry out photosynthesis. The algae provide the corals with important nutrients. In fact, without their algal partners, the corals eventually die. Whenever the seawater around a coral reef gets too warm, however, corals expel the algae from their bodies. Global warming could increase the temperature of ocean waters enough so that coral reefs in many parts of the world might expel their algae and die as a result.

Even if corals survive warmer water

temperatures, they will still be in danger from rising sea levels. Reef-building corals cannot live in more than about 100 feet (30 m) of water because the algae they contain need sunlight to carry out photosynthesis. But corals grow upward by only a fraction of an inch each year. If sea levels rise as rapidly as researchers predict they will, many of the world's coral reefs will not be able to grow fast enough to keep up with the rising water. The reefs will "drown" as sea levels go up, and the many organisms that live on or around them probably will die along with them.

Wetlands

Wetlands are places where land and water meet. Where salty oceans lap the shore, we find coastal wetlands such as estuaries, salt marshes, and mangrove swamps. Inland there are freshwater wetlands—marshes, swamps, bogs, and prairie potholes. Wetlands are home to all sorts of small animals and are ideal nesting spots for ducks, geese, and other birds. Coastal wetlands form "nurseries" where fish come to spawn and where shellfish live in great numbers. Some species of plants are able to grow only in the special kind of habitat that wetlands provide.

Rising sea levels are expected to destroy many of the world's coastal wetland areas. Up until recently, most wetlands have been able to keep pace with slowly rising sea levels. As the water has advanced, wetlands have gradually moved inland. But scientists predict that it won't be long before sea levels will be rising too fast for wetlands to keep up. It is estimated, for example, that a three-foot (1-meter) rise in sea level will destroy up to two-thirds of the wetlands now found along the coasts of the United States.

Freshwater wetlands will suffer from rising temperatures and a changing climate. Higher temperatures could harm fish in ponds and small streams. And droughts will dry up potholes, swamps, and marshes. This would leave all kinds of animals, especially birds, with fewer places to live and to raise their young.

Animals

Most biologists think that a rapidly changing climate will be harder on plants than on animals. Part of the reason for this is that changes in temperature and rainfall affect plants very quickly. Few kinds of plants can cope with stressful conditions for very long. Many kinds of

Freshwater wetlands like this Minnesota marsh could be dried up by the drought that would accompany global warming in some regions.

animals, on the other hand, do have ways of surviving periods of environmental stress, at least for a while.

When times are bad, for example, many insects go into a kind of "suspended animation" and remain inactive for long periods. Birds tend to lay fewer eggs so there will be fewer mouths to feed. Female kangaroos can even delay the development of embryos they are carrying until the weather or food supplies improve.

But the changes in temperature and climate brought about by the greenhouse effect will be irreversible—conditions may never return to "normal." As the years pass, if animals continue to produce fewer and fewer offspring, or don't reproduce at all, their populations gradually will decrease. Eventually, some species may become extinct as a result.

Climate change could also affect the time of year when some animals reproduce. And whenever the timing of reproduction in one species changes, it can affect the survival of other species. The journeys of migrating birds, for instance, are timed to correspond with the availability of food along the way and at their final destination. After spending the winter in South and Central America, dozens of different species of North American birds fly all the way up to the arctic tundra each spring to lay eggs and raise their young.

The chicks hatch at the same time that insects are emerging by the millions from the shallow ponds that dot the tundra. If temperatures were to warm up earlier each spring because of global warming, the insects might emerge before the birds arrived. By the time nests were built and eggs were laid and hatched, most of the birds' food supply could be gone.

The fact that animals are able to move from place to place does give them a better chance of surviving climate changes. Unfortunately, in today's world there are few places where animals on the move will not find natural or human-made barriers in their path. Mountain ranges and urban developments could prevent many species from completing their journey to a new habitat.

At this point, no one can predict which animals will be winners or losers in this large-scale reshuffling of life on earth. But it is not too early to see that some species are more at risk than others.

Conservationists are very worried, for example, about the future of animals that are now confined to wildlife reserves around the world. When many reserves were set up years ago, no one anticipated that the world's climate would change. What will the animals now living in reserves do when their habitat changes as a result of the greenhouse effect? In order to survive they will be forced to move away from the protected areas that have been set aside for them. Once outside the reserves, they will be facing a hostile environment.

What is even more tragic is that many of the animals living in wildlife reserves are already endangered. Enormous efforts have been made to save elephants, pandas, cheetahs, tigers, rhinos, mountain gorillas, and many other animals from extinction. Now, because of the greenhouse effect and the impact it will have on our planet, these and many more animals may vanish from

Changes in habitat caused by global warming may cause elephants and other endangered animals to become extinct.

the face of the earth in your lifetime.

It's hard to imagine a world without elephants and coral reefs, without polar bears and arctic tundra. It's frightening to think about the world's food reserves dwindling away or entire islands disappearing under rising seas. Yet this is what scientists predict our world could be like in the next century if greenhouse gases continue to build up in the atmosphere at the rate they are now.

Does the future have to be this bleak? Or is there something we can do about the greenhouse effect and global warming?

83

Workers planting pine trees in a U. S. national forest in Mississippi. Reforestation is one of the steps that can be taken to reduce carbon dioxide in the atmosphere.

6

WHAT CAN WE DO?

What can we do about the greenhouse effect? On one hand, of course, we could do just what we have been doing. We could continue to burn huge amounts of fossil fuel, to use CFCs, and to destroy forests, and simply hope that the predictions made by scientists are wrong. There are some people who have this "wait and see" attitude. They believe that nothing should be done until the predictions that scientists have made are proven right or wrong, beyond a shadow of a doubt.

It will probably be several years, perhaps even another decade, before we know just how accurate the predictions that have been made about the greenhouse effect really are. But can we afford to wait that long before doing something about the problem? Almost all scientists who study the greenhouse effect agree that the longer we wait, the more greenhouse gases will accumulate in the air and the greater the problem we will have to deal with in the future. They believe that the time to "wait and see" is past. It is time to act. But what can we do?

A number of people have suggested different ways in which we could alter the environment to try to "counteract" the greenhouse effect. For example, by fertilizing the oceans with chemicals, it might be possible to cause a population explosion among photosynthesizing plankton. If there were more of these plankton in the sea, they might be able to remove more carbon dioxide from the environment.

Another idea is to prevent some of the sun's energy from reaching the earth. This could help cool the earth and reduce global warming. By releasing planeloads of dust into the atmosphere or sending giant sunshields into orbit around the planet, we might be able to block sunlight on a global scale. Perhaps the warming trend could also be slowed by taking carbon dioxide and other greenhouse gases out of the air. Some people suggest setting up huge fans that would draw air into treatment plants where carbon dioxide could be removed from it. Others suggest constructing super-lasers that would sweep through the atmosphere and destroy molecules of synthetic chemicals such as CFCs.

As intriguing as some of these ideas may be, they are not very practical suggestions for dealing with the greenhouse effect. First of all, at the moment we do not have the technology to build sunscreens in space or CFC-destroying lasers. Second, most of these proposals would cost a great deal of money, and there is no guarantee that they would work. And, finally, these "solutions" to the greenhouse effect might cause more problems than they solve. Who could predict the effects of dumping vast amounts of fertilizers into the sea? Where would we put carbon dioxide if we could extract it from the air—bury it? Dump it into the ocean? And what if we put too much dust into the atmosphere? Would the earth cool too fast and enter another Ice Age? Tampering with the environment more than we already have could create a situation on earth that is even worse than the greenhouse effect!

So where does that leave us? At this point, the best thing we can do to control the greenhouse effect is to drastically reduce the amounts of heat-trapping gases being added to the atmosphere. Even this won't solve the problem completely. There will be some amount of global warming no matter what we do because of the gases that have already accumulated in the air. But by reducing greenhouse gases now entering the atmosphere, we can at least slow down the warming that is going to take place in our global greenhouse. Reducing greenhouse gases is not an easy task, however. It is a job that is going to require the willingness and determination of people all over the world.

The smog that pollutes the air over Los Angeles is produced by burning large quantities of fossil fuels.

REDUCING CARBON DIOXIDE

In order to slow global warming, scientists estimate that we need to cut in half the amount of carbon dioxide that we now release into the air each year. There are at least three ways in which we can do this—by conserving energy, by reforesting the earth, and by switching to renewable and alternative forms of energy.

Energy Conservation

A large portion of the energy we use comes from burning fossil fuels. Most power plants burn fossil fuels, especially coal, to generate electricity. Both natural gas and oil are widely used for heating homes, schools, and offices. Cars, trucks, motorcycles, and lawnmowers all burn gasoline, which is extracted from oil. Each time we turn on the lights, turn up the

87

thermostat, or start our cars, we use energy that comes from fossil fuels. And each time we use this energy, carbon dioxide is directly or indirectly released into the air. The average car, for instance, adds at least its own weight in carbon dioxide to the atmosphere each year. The average coal-fired power plant adds more than a million tons of carbon dioxide to the air annually.

Energy conservation is one of the quickest and easiest ways to reduce the buildup of carbon dioxide in the atmosphere. If we really want to control the greenhouse effect and global warming, then people everywhere must learn to use energy more wisely. This is especially true in the United States. The U.S. consumes more energy than any other nation on earth and contributes one-fifth of all the CO_2 that enters the atmosphere each year.

Energy conservation is nothing new in the United States. During the 1970s, oil prices rose sharply. Suddenly people had to pay a lot more to drive their cars, heat their homes, and operate their appliances. It wasn't long before energy conservation became a way of life. Automobile manufacturers built smaller, more fuel-efficient cars. New houses were well insulated.

Across the country, public buildings and industries were made more energy-efficient. And it all made a difference. Between 1973 and 1986, energy conservation saved the United States roughly 13 million barrels of oil each day!

In the late 1980s, however, the price of oil fell, and many people forgot all about energy conservation. Automobile manufacturers started making bigger, less efficient cars. New buildings were not constructed to use energy as efficiently as they could. All in all, people in the United States are using more energy now than ever before. Yet because of the greenhouse effect, it has never been more important to conserve energy than it is today. Energy conservation is no longer just a matter of saving money—it could very well be a matter of saving our planet. And energy conservation is one way in which every person can help slow the rate of global warming and climate change.

Conserving energy means making changes in the way we live. It means walking or biking instead of taking a car. Or, if you must ride, taking the bus or some other form of public transportation whenever possible. Conserving energy means turning off the lights, TV, and stereo

Leaving your car at home and taking public transportation can help to cut down on carbon dioxide buildup.

paper, glass, plastic, and aluminum—it takes energy to make all of these products. Talk to your family about ways in which all of you could save energy. When it comes to energy conservation, little things can mean a lot. By replacing a single 75-watt incandescent light bulb with an 18-watt fluorescent bulb that is just as bright, at least 300 pounds of coal can be saved.

If you are in school, ask your teachers about saving energy. Perhaps you and your classmates could design a class project to save energy at school. Write a letter to your congressional representatives and tell them you are worried about the greenhouse effect and global warming. These people are making decisions today that will affect the world you will be living in tomorrow.

Some of the changes we need to make in our lives in order to save energy may not be pleasant. It is hard to cut down on conveniences that we are accustomed to using all the time. In order to get people to conserve energy effectively, it may become necessary for governments to provide incentives to encourage them to do so.

For example, cars and other vehicles that use a lot of fuel may need to be heavily

when you leave the room. It means turning down the thermostat in your house in the winter and turning it up in the summer. Or better yet, try using fans instead of air conditioning in the summer months.

You can help save energy by recycling

The development of more fuel-efficient automobiles could play a role in energy conservation. The experimental Volvo shown here can travel 65 miles (about 104 kilometers) on one gallon of fuel.

taxed. In the United States, some politicians have suggested adding at least a dollar-a-gallon tax on gasoline. They believe that this action would encourage people to drive less often, use public transportation more often, and buy more fuel-efficient cars. Rebates and tax credits for doing such things as insulating houses and office buildings and replacing old appliances with new, energy-saving models may also

be useful in encouraging conservation.

Serious efforts to conserve energy will have many benefits. Most importantly, energy conservation will reduce carbon dioxide in the atmosphere and slow the rate of global warming. It will also reduce air pollution and improve the quality of the air we breathe. Energy conservation would mean that we would not need to dig as many coal mines or import as much oil from foreign countries as we do now. And energy conservation, in the United States and elsewhere, will also buy the world time in which to discover and develop new sources of energy to replace fossil fuels.

Reforestation

Another thing we can do to help reduce the amount of carbon dioxide that accumulates in the air is to plant millions of new trees. This process is known as **reforestation**.

Like other green plants, trees use carbon dioxide during photosynthesis. Trees can take up a surprising amount of CO_2 from the atmosphere. An acre of sycamore trees, for example, can absorb about 12 tons of carbon dioxide each year. An entire forest of sycamore trees can absorb many times more.

But remember that burning fossil fuels adds about 20 *billion* tons of carbon dioxide to the air each year. In order to remove that much carbon dioxide from the air each year, we would need to plant enough trees to cover an area roughly the size of Australia! It's not very likely that we will be able to do that. But reforestation on a somewhat smaller scale can at least help slow the buildup of carbon dioxide in the air and delay global warming.

Some industries are using reforestation as a way to make up for the carbon dioxide they are releasing into the air. One U.S. power company, for example, is paying to have 52 million trees planted in Guatemala while it is building a new coal-fired power plant in Connecticut. The forest of trees is expected to absorb about the same amount of carbon dioxide each year that the power plant will release into the atmosphere.

Like conserving energy, planting trees is a way in which many people can become involved in helping to control the greenhouse effect and global warming. Is there a place in your yard or neighborhood where a few trees could be planted? If you can plant trees where they will eventually shade your house, apartment building, or school, you'll be helping to conserve energy, too. Tree-shaded buildings stay cooler in the summer, so there is not as much need for air conditioning.

Reforestation is certainly not a "cure" for the greenhouse effect. But like energy conservation, it will help control the problem while we are developing new energy sources that do not add carbon dioxide to the atmosphere.

These solar collectors on the roof of a building can help to provide heat and hot water. In the future, other solar devices may be used for the large-scale production of electricity.

New Energy Sources

There is no getting around the fact that our world needs energy. But we cannot go on using fossil fuels as we are now without doing more and more damage to the earth and its climate. Therefore it is very important that the world quickly find new energy sources that do not add carbon dioxide (or other greenhouse gases) to the atmosphere.

Solar, wind, and geothermal power are ideal choices for new energy sources.

Energy harnessed from the sun, from winds that sweep across the landscape, and from heat stored deep within the earth can be used to make electricity. All these forms of energy are clean, safe, and renewable—they can never be used up like fossil fuels eventually will be.

Unfortunately, the world cannot easily switch to using these renewable energy sources at this point. The technologies of solar, wind, and geothermal power have problems that must be solved before they can be used for large-scale production of electricity. Cost is also a problem. Electricity produced by solar devices, for example, currently costs about five times more than electricity produced by coal-fired power plants.

Eventually, advances in technology should lower the cost of solar power and should also make other renewable energy sources more practical replacements for fossil fuels. But technological advances come from research, and research costs money. Nations around the world will have to be willing to spend what it takes to develop these new sources of power quickly.

Nuclear power is another option. In the process of generating electricity, nuclear power plants do not release carbon dioxide into the atmosphere. But should power plants that burn fossil fuels be replaced by nuclear reactors simply because they don't release CO_2? Many people don't think so because they worry about the safety of nuclear power plants. They also point out that nuclear plants are expensive to build and operate. Furthermore, nuclear power plants produce dangerous radioactive wastes that must be safely stored for many years. In the long run, money spent building nuclear power plants might be better invested in energy conservation or in developing safe, renewable energy sources such as solar power.

It is very likely that solar, wind, and geothermal power can eventually be used to produce the electricity we need for many things. But what about energy needed to operate our cars and other vehicles that now run on fossil fuels? One potential source of energy for transportation is alcohol fuel.

Alcohol fuel is produced from plant materials such as corn. Corn kernels are first ground up, boiled, mixed with water, yeast, and other ingredients, and then put into large tanks. The mixture ferments, and the alcohol that is produced during the fermenting process is drawn off and purified into fuel.

Alcohol fuel gives off few pollutants as it burns. And although carbon dioxide is released in the burning process, alcohol fuel does not add any *extra* carbon dioxide to the atmosphere in the way that fossil fuels do. This is because the plants used to make alcohol fuel take carbon dioxide out of the atmosphere while they are growing. Thus, there is a continual balance between the amount of carbon dioxide removed from the air by the corn plants and the amount that enters the air when the fuel made from the plants is burned. This balance does not exist when fossil fuels are burned because the plants that produced these fuels lived and died millions of years ago. As a result, burning fossil fuels only adds carbon dioxide to our present-day atmosphere.

REDUCING OTHER GREENHOUSE GASES

As you know, carbon dioxide is not the only greenhouse gas. In order to slow down the rate of global warming, something also has to be done to reduce the other greenhouse gases that are being added to the atmosphere. This will be harder to do for some gases than for others.

It is going to be very difficult, for example, to solve the methane problem. This is because most of the methane rising up into the atmosphere is being produced by natural processes that we cannot control. We could try to trap the methane that escapes from coal mines, natural gas pipelines, and landfills. But how can we trap gas rising from millions of acres of rice paddies?

How can we stop cows from belching and termites from digesting their food? And there is nothing we can do about the enormous quantities of methane gas that researchers expect will be released from the arctic tundra if it thaws out.

We may be able to do more about nitrous oxide. Some of the nitrous oxide being released into the atmosphere comes from the burning of fossil fuels. If our cars and factories burned fossil fuels more efficiently, the amount of N_2O in the air might not increase as fast as it does now. It also would help if farmers could cut back on their use of nitrogen fertilizers and if deforestation was brought to an end worldwide. But we will probably never be able to control completely the amount of nitrous oxide that rises into the atmosphere each year.

Fortunately, we have a much better chance of solving the problem of CFCs. Chlorofluorocarbons are synthetic compounds that only we humans produce and release into the environment. Because we make CFCs, we should be able to limit the use of these destructive substances immediately and eventually stop producing them altogether.

What can we do now to reduce the

Old refrigerators contain chlorofluorocarbons that can leak out. Removing the CFCs before the appliances are discarded can keep these dangerous chemicals out of the atmosphere.

amount of CFCs in the atmosphere? Although the United States and several other countries have already banned CFCs in almost all of their aerosol products, aerosol cans are still the world's largest source of CFCs. Therefore, one of the first steps that should be taken is to ban the use of CFCs in all aerosol cans everywhere in the world.

Large amounts of chlorofluorocarbons are released into the environment each year from air conditioners and refrigerators. When the coolant in a car's air conditioner is recharged, for example, some of this CFC-containing fluid usually leaks out. Spilled coolant is often washed away or simply allowed to evaporate. Whenever possible, such coolants should be carefully recovered and reused.

Old refrigerators and other appliances that contain CFCs often end up in junkyards and garbage dumps. They eventually

95

fall apart, and their CFCs leak out, evaporate, and rise up into the atmosphere. Several countries are working on a solution to this problem. In West Germany, for example, old refrigerators are now collected and their CFCs carefully removed before they are discarded.

In the manufacture of plastic foam products, CFCs are used to inflate the foam and form it into various shapes. Huge quantities of CFCs escape into the air during this process. By capturing these chemicals and using them over and over again, plastic foam manufacturers could greatly reduce the amount of CFCs being released into the atmosphere. And they could save money at the same time.

Unfortunately, simply controlling the amounts of CFCs that escape into the environment won't be enough. The world cannot afford to produce any more CFCs that act as greenhouse gases and destroy the ozone layer. Researchers are now searching for new chemicals that can do the same jobs as the CFCs we use now, but without damaging the environment. Several possible substitutes for some of the most widely used chlorofluorocarbons have already been found. CFC-113, for example, may soon be replaced by a

A scientist at a DuPont laboratory tests a material developed to replace CFCs in cleaning electronic circuit boards.

substance that is made from the rinds of citrus fruits. This new chemical compound cleans computer chips just as well as CFC-113 and does not seem to damage the environment.

PLANNING FOR CLIMATE CHANGE

If we can reduce the greenhouse gases going into the atmosphere, we probably can slow the rate of global warming and climate change. But sooner or later the world will feel the impact of the greenhouse effect. There are a number of things that we can do to prepare for the changes that are coming. If we act now, perhaps we can "soften the blow" of the greenhouse effect.

Drought could soon become a more frequent problem in the world's grain-growing regions. Farmers in these areas need new varieties of wheat, corn, and soybeans that will be more resistant to drought and heat and still produce good yields. Farmers may also need to plant a wider variety of crops. In the future, the weather could change much more dramatically from year to year than it does now. The more different types of crops farmers

Storing grain for future use is one way of planning for climate change.

97

A researcher at the U. S. Department of Agriculture examines thriving lima bean plants that have been developed to withstand drought. In the foreground are the withered stalks of corn plants that are not drought-resistant.

raise, the better the chance that at least some will survive unpredictable growing conditions.

Drought usually creates water shortages. By designing new kinds of irrigation systems that waste less water, farmers in some regions might be able to continue raising crops even if droughts become more frequent and severe because of the greenhouse effect. And it is not too soon for both small towns and large cities to begin planning for possible water shortages

in the future. It would be better to build new wells and reservoirs now than to wait until a particularly dry year when water is in short supply.

Because of global warming and climate change, agricultural production could be very unpredictable in some parts of the world for many years to come. In order to avoid famine, the world should have enough surplus grain on hand so that food can be sent to countries where crops have failed. Since the mid-1980s, however, world supplies of stored grain have become dangerously low. Our world will be a safer place if these food reserves can quickly be built back up again.

What about wild animals? What will we need to do to help them survive changes caused by the greenhouse effect? As global warming alters habitats, many kinds of animals will be on the move, but all sorts of barriers will stand between them and a new place to live. To help animals get around these barriers, it might be necessary to set up **migration corridors** that connect natural areas with one another. Animals could then travel along these "pathways" to reach new habitats.

The United States already has made plans for a nationwide system of these corridors that will connect parks and other natural areas all over the country. In many other parts of the world, however, migration corridors may need to cross national borders. Getting neighboring countries to agree on such large-scale projects could be very difficult.

Not even a worldwide system of migration corridors, however, will be enough for some animals. Wildlife managers may have to capture and move certain species to new habitats. Even this kind of help will not guarantee their survival.

Without human intervention, many kinds of plants also may not survive as the earth's climate changes. Forests, in particular, may need our help. If climate changes come rapidly, few tree species will be able to spread into new areas fast enough to keep up with changing conditions. It may be necessary to transplant trees into new habitats in order to save them. Transplanting trees, however, is not the same as transplanting a forest, complete with all its plant and animal inhabitants. Whatever we do, we will never be able to recreate the natural systems that will be lost to the greenhouse effect.

WORKING TOGETHER

The greenhouse effect will affect all life on earth. It is a global problem, one that must be faced by people everywhere. No nation acting alone will be able to slow the pace of global warming and climate change. Our only hope of dealing with this enormous environmental problem is international cooperation. But getting countries to work together is a very challenging task.

To a large extent, the industrialized nations of the world have caused the problems associated with the greenhouse effect. By burning great quantities of fossil fuels over the last two centuries, these nations have set the process of global warming in motion. And they still are the largest contributors to the problem. Today, industrialized nations consume four-fifths of all the fossil fuels burned each year.

It is hardly surprising, then, that people in many developing nations believe the responsiblity for solving the greenhouse problem rests on the shoulders of those who played the largest role in creating it. They feel particularly strongly about this because climate models predict that developing countries could suffer the most from drought, rising sea levels, and other devastating effects of global warming.

To make matters worse, people in most developing nations are trying to achieve a higher standard of living. They want to enjoy the same benefits of modern society that people living in industrialized countries now take for granted. But modernization and industrialization require a great deal of energy. Just when it has become so important for the world to reduce its consumption of fossil fuels, many develop-

ing countries are beginning to use more coal, oil, and natural gas than ever before. If this trend continues, experts predict that by the year 2005, developing countries could be producing more carbon dioxide than industrialized countries do today. It will be difficult for people in developing nations to choose to limit their use of fossil fuels and in so doing possibly give up a chance for a better way of life.

And is it really fair to expect them to do so? The United States and other industrialized countries have achieved their current high living standards largely by being able to use cheap, readily available energy. Do we have the right to deny others the same opportunity? Perhaps industrialized nations should contribute some of their resources to improve living conditions for people in developing countries. With such assistance, global fossil fuel consumption might be kept in check while new energy sources were developed. In light of the seriousness of the greenhouse effect and global warming, shouldn't we be willing to make such a contribution toward safeguarding our planet?

But how far should we be willing to go? How many economic and personal sacri-

These Chinese farmers are using a foot-operated device to irrigate their fields. Their lives would be easier if a machine could do the same job, but if the machine ran on fossil fuels, its use would increase the amount of carbon dioxide in the atmosphere.

fices will people and nations be willing to make for the "good of the earth"? How will they be convinced to make these sacrifices? Who will decide what actions need to be taken? And how will these decisions be enforced? Obviously, the greenhouse effect is an enormously complicated problem. There will be no simple solutions.

Some progress, however, is being made in bringing the world together on the greenhouse issue. A growing number of international meetings about the greenhouse effect are held each year. Scientists and political leaders from around the world come to these meetings to share new information about the earth's natural

Only by working together can we hope to solve the greenhouse problem.

systems. They study the most current predictions about global warming and its impact on the earth. And they discuss possible ways to deal with the problems that climate change will bring. But meet-ings alone won't solve the greenhouse problem. It will take action by individuals, by governments, and by the whole world working together.

Over the past 200 years, human beings

have brought about enormous changes in the environment. In our efforts to improve the quality of our lives, we have polluted the waters, the land, and the atmosphere. In many cases, we were not fully aware of the consequences of our actions. Even when we were aware, we often went ahead with the attitude that the earth would somehow recover from all the terrible things we were doing to it.

But now it seems that our actions have finally caught up with us. We have irreversibly changed the nature of the atmosphere—our planet's life-support system. The greenhouse effect is a very real problem. As we have seen, it may be the most serious environmental problem that humans have ever faced. But we must face it if we want to preserve life on our planet as it exists today. The earth is our home, our only home in the vast emptiness of space. Can we afford to destroy it?

GLOSSARY

carbon dioxide (CO_2)—a colorless, odorless gas that occurs naturally in the atmosphere and plays an important role in the lives of plants and animals. Plants take in CO_2 during the food-making process of photosynthesis. Animals release the gas during the process of respiration. Photosynthesis and respiration normally maintain a fairly constant level of CO_2 in the atmosphere, but the burning of fossil fuels and the destruction of forests have upset this balance. As a result, carbon dioxide is building up in the atmosphere and trapping more of the heat energy radiating from the earth's surface. Currently, CO_2 is the most important greenhouse gas.

chlorofluorocarbons (CFCs)—synthetic chemicals made up of different combinations of chlorine, fluorine, and carbon. Used for many industrial purposes, CFCs contribute to the greenhouse effect by trapping heat energy radiated from the earth's surface. They also destroy the ozone that protects the earth from the sun's harmful ultraviolet radiation.

climate models—complex computer programs that simulate the operation of the earth's climate. By changing different parts of a program—for example, the amount of methane in the atmosphere—scientists can predict the effect of that change on temperature and other variables.

deforestation—the destruction of the world's forests to clear land for agriculture or to provide timber. Deforestation eliminates trees that absorb carbon dioxide. If the trees are burned, large amounts of CO_2 are added to the atmosphere.

electromagnetic radiation—energy that travels through space in the form of waves. Energy given off by the sun is electromagnetic radiation.

electromagnetic spectrum—the range of electromagnetic radiation, made up of energy waves with different wavelengths

fossil fuels—fuels like coal, oil, and natural gas that were formed from the remains of plant material deposited during the earth's Carboniferous period

general circulation models (GCMs)—the types of climate models that scientists use to study the greenhouse effect. These very detailed models include large amounts of information about the earth's atmosphere, the oceans, and world climate.

greenhouse effect—the process by which gases in the earth's atmosphere absorb heat energy radiating from the surface and then reradiate some of that heat back toward earth. As a result, a certain amount of heat energy is prevented from escaping immediately into space. This "trapped" heat keeps the planet's surface comfortably warm, but as the concentrations of greenhouse gases in the atmosphere increase, the greenhouse effect is strengthened and the average global temperature rises.

greenhouse gases—gases like carbon dioxide, methane, and nitrous oxide that "trap" heat in the earth's atmosphere

global warming—an increase in the average temperature near the earth's surface caused by the strengthening of the greenhouse effect. Some scientists predict that the average global temperature may rise as much as 9° F (5° C) by the year 2050.

halons—synthetic chemicals widely used in fire extinguishers. Like CFCs, halons act as greenhouse gases in the atmosphere.

ice cores—long, cylindrical pieces of ice taken from ice sheets and glaciers. By studying the gases trapped in these ice samples, scientists can learn something about the earth's atmosphere in past periods.

infrared radiation—waves of radiant energy with wavelengths from approximately 0.7 to 100 micrometers on the electromagnetic spectrum. Infrared radiation is called heat energy because we can feel it as heat.

mesosphere—the third layer of the earth's atmosphere, extending to about 50 miles (80 km) above the surface

methane (CH_4)—a colorless, odorless, flammable gas that is produced by a variety of natural sources, including microorganisms in waterlogged soil and in the digestive tracts of cattle and termites. After carbon dioxide, methane is the most important of the greenhouse gases.

migration corridors—pathways that might have to be created to connect natural areas such as parks and games reserves with one another. Animals could use these corridors to move to new living places if their habitats are altered by global warming.

nitrous oxide (N_2O)—a colorless gas with a slightly sweet odor. Also known as laughing gas, N_2O comes from the burning of fossil fuels, the use of nitrogen fertilizers, and other sources. In the atmosphere, it acts as a greenhouse gas.

ozone layer—a layer of ozone gas in the stratosphere that shields the earth from most of the ultraviolet radiation coming from the sun

parts per million (ppm)—units used to measure atmospheric gases such as carbon dioxide. In this system of measurement, the concentration of a gas is expressed in terms of the number of its molecules contained in one million molecules of air.

peat—organic material made up of partially decayed plants. Fossil fuels were formed from peat buried in the ground millions of years ago. Today, large amounts of peat are contained in the soil of the arctic tundra.

photosynthesis—the process by which green plants use energy from sunlight to produce their own food. As part of this process, plants take in carbon dioxide and water, and release oxygen.

reforestation—the planting of large numbers of new trees to take the place of forests that have been cut down

respiration—the chemical process in which energy is released when organic compounds are broken down using oxygen. Carbon dioxide is produced during respiration.

solar energy—energy produced by explosive nuclear reactions that take place in the interior of the sun. Radiating from the sun's surface, solar energy travels through space in the form of waves of electromagnetic radiation.

stratosphere—the second layer of the earth's atmosphere, extending out to about 30 miles (48 km) above the surface. The ozone layer is contained within this part of the atmosphere.

thermal expansion—an increase in the size of a solid or the volume of a liquid due to heating. When water is heated, the liquid expands and takes up more space. As global warming occurs, thermal expansion will cause sea levels to rise.

thermosphere—the fourth and outermost layer of the earth's atmosphere, extending out about 600 miles (965 km) above the surface

trace gases—gases such as hydrogen, helium, and neon that are present in very small amounts in the atmosphere

troposphere—the innermost layer of the earth's atmosphere, extending out about 7 miles (11.3 km) above the surface

ultraviolet (UV) radiation—waves of radiant energy with wavelengths ranging from about 0.0005 to 0.4 micrometers on the electromagnetic spectrum. Ultraviolet waves coming from the sun usually have wavelengths between 0.2 and 0.4 micrometers. Most of this high-energy radiation is absorbed by the ozone layer in the stratosphere.

visible light—waves of radiant energy with wavelengths ranging from about 0.4 to 0.7 micrometers on the electromagnetic spectrum. We can see this form of electromagnetic radiation because our eyes are sensitive to these wavelengths.

water vapor—water in the form of an invisible gas. Like greenhouse gases, water vapor in the earth's atmosphere absorbs some of the heat energy radiated from the surface of the planet.

wavelength—the distance from one high point of a wave to the next high point (or from one low point to the next low point). Waves of electromagnetic radiation are measured in terms of their wavelengths.

INDEX

Researchers at NASA's Goddard Institute for Space Studies study a computer image containing information that will be used in climate modeling.

ACKNOWLEDGMENTS The photographs and illustrations in this book are reproduced through the courtesy of: p. 6, USDA; p. 9, Bachman's Garden Center; pp. 11, 60, 63, NASA/Goddard Institute for Space Studies; pp. 12, 24, 104, NASA; p. 26, Husky Oil Corporation; p. 31, Library of Congress; pp. 32, 102, Wisconsin Department of Natural Resources; p. 34, © Richard O. Bierregaard; p. 36, Julie Palais; p. 37, Anthony J. Gow, Cold Regions Research and Engineering Laboratory, Department of the Army; pp. 38, 40, Scripps Institute of Oceanography; p. 42, Rick Hansen, Minnesota Department of Agriculture; p. 44, Janda Thompson; p. 45, Ceylon Tourist Board; p. 47, S. A. Johnson; p. 49, NASA/Goddard Space Flight Center; p. 54, Winzen International, Inc.; p. 57, National Center for Atmospheric Research; p. 59, V. Ramanathan, University of Chicago; p. 61, Jay A. Stravers; pp. 64, 75, 81, Minnesota Department of Natural Resources; p. 66, Tony Craddock/Tony Stone Worldwide; p. 68, John Deere Company; p. 70, Florida Department of Commerce; p. 71, Agency for International Development; p. 73, Breck P. Kent; p. 77, Alaska Division of Tourism; p. 78, Industry, Science and Technology Canada Photo; p. 83, Independent Picture Service; p. 84, USDA Forest Service; p. 87, Los Angeles County Pollution Control District; p. 89, Metropolitan Transit Commission, Minneapolis; p. 90, Volvo North America Corporation; p. 92, Solar Energy Research Institute; p. 95, Jerry Boucher; p. 96, DuPont Company; p. 97, Monsanto Company; p. 98, USDA Agricultural Research Service; p. 101, F. Botts/FAO.